Characterisation Techniques for Civil Engineers

The primary aim of this book is to provide an understanding of the sophisticated, modern characterisation techniques in the domain of civil engineering. It systematically covers physical, chemical, mineralogical and microstructural characterisation, which is imperative to evaluate the construction materials and their performance. It describes tools such as rheometers, thermogravimetric analysers, scanning electron microscopes, X-ray diffractometers and other miscellaneous methods. In each chapter, a detailed scientific background, instrumentation details, working principles, and applications of a specific technique are provided.

Features:

- Describes rheological and microstructural characterisation testing.
- Discusses sophisticated characterisation techniques for construction materials.
- Explains the detailed procedure of sample preparation and testing.
- Provides detailed descriptions of different parts of the instruments and their purposes.
- Includes questions and answers at the end of each chapter.

This book is aimed at graduate students and researchers in civil engineering.

Characterisation Techniques for Civil Engineers

Bahurudeen A, Rithuparna R, and P V P Moorthi

CRC Press
Taylor & Francis Group
Boca Raton London New York

CRC Press is an imprint of the
Taylor & Francis Group, an **informa** business

Designed cover image: © Shutterstock

First edition published 2024
by CRC Press
2385 NW Executive Center Drive, Suite 320, Boca Raton FL 33431

and by CRC Press
4 Park Square, Milton Park, Abingdon, Oxon, OX14 4RN

CRC Press is an imprint of Taylor & Francis Group, LLC

ISBN: 9781032555423 (hbk)
ISBN: 9781032635385 (pbk)
ISBN: 9781032635392 (ebk)

DOI: 10.1201/9781032635392

Typeset in Times
by codeMantra

The authors dedicate this book to their parents

Abdulsalam and Habibunnisha

Rajesh and Beena

Subbiah and Palaniammal

Contents

Preface

Demand for construction materials has notably increased because of infrastructure development. One of the major challenges in the construction sector is finding good-quality construction materials. For instance, in many developing countries, using river sand as fine aggregates in concrete has been banned to avoid excessive riverbed mining and protect the riverbed. Hence, there is a significant need for alternative fine aggregates in the construction sector. Physical, chemical, mineralogical and microstructural characterisation of construction materials is vital to understanding the potential of any alternative material for use in construction. In addition to conventional testing methods, most international standards include the use of characterisation techniques in the selection and performance assessment of construction materials. Hence, the present need is to understand the use of characterisation techniques, specifically in the quality control of civil engineering materials. Therefore, this book focuses on the technical details of sophisticated characterisation techniques used for construction material testing in a simple way with high precision. This book is divided into five chapters. The first chapter focuses on rheological measurements of construction materials. Chapters 2–4 describe particular widely used characterisation techniques. Chapter 5 explains miscellaneous characterisation techniques used in the quality control of construction materials. Besides, out-of-date and unneeded details are completely sidestepped in this book. Additionally, neat illustrations are given for easy understanding of readers throughout the book. At the end of each chapter, practice questions and answers are presented. The authors sincerely express gratitude to the editorial team of CRC Press for their continuous support, and they welcome feedback from readers.

Bahurudeen A, Rithuparna R, and P V P Moorthi

About the Authors

Bahurudeen A is an Associate Professor in the Civil Engineering Department at Birla Institute of Technology and Science (BITS Pilani), Hyderabad Campus, India. He completed his PhD in Civil Engineering at Indian Institute of Technology Madras (IIT Madras). He served as a Senior Scientific Officer at IIT Madras. His research focus involves a scientific understanding of properties of construction materials. He has published seven patents and several research articles in reputed journals and conferences on the characterisation of construction materials. He has published a book titled *Testing of Construction Materials*, with CRC Press as an international edition. He has completed several real-time consultancy projects and is actively involved in many educational bodies.

Rithuparna R is the Head of Academic Instructions and Scheduling Division of Engineering Delight Academy, Salem, India. She completed her Master's in Structural Engineering at Birla Institute of Technology and Science (BITS Pilani), Hyderabad Campus. She published several research articles in reputed journals and conferences. Her research interests include sustainable alternative building materials, characterisation of supplementary cementitious materials and durability assessment of structures.

P V P Moorthi is the Technical Head of Engineering Delight Academy, Salem, India. He completed his PhD in Civil Engineering at Indian Institute of Technology Bombay (IIT Bombay) and his Master's in Infrastructure Engineering and Management at Birla Institute of Technology and Science (BITS Pilani), Rajasthan. He has published several research articles in reputed journals and conferences. His areas of exposure include construction materials, rheology of cementitious materials and non-destructive testing.

1 Rheology

1.1 INTRODUCTION

Rheo-physical investigation involves the study of material deformation. Especially, it sets out a basis for classifying the material qualitatively and quantitatively under flow. Mixing fluids with elements (solids, semi-solids, etc.) results in complex suspensions that range from a perfect liquid to an elastic solid. Suspensions that consists of solid particles immersed in a liquid form the simplest material. Based on the physical, chemical, and surface characteristics of the particles, the behaviour of the suspension varies. Particles in the suspension undergo different types of forces at rest and shear. The balance between the forces determines the behaviour and nature of the suspension. At rest, depending on the characteristics of the suspension, attractive and repulsive forces persist. Attractive forces are attributed to van der Waals' forces, hydrogen bonding due to dipole–dipole interaction, ion-correlation forces arising due to the reactivity of the particles in the suspension, etc. Similarly, repulsive forces of the particles in the suspension are due to double layer formation that results in high zeta potential, electrostatic repulsion, and steric hindrance. For the suspensions to be in a stable state, the repulsive forces of the particles in the suspension should be higher than the attractive forces. This results in a dispersed and moderately stable suspension (higher zeta potential, typically more than ±30 mV).

On the other hand, if the zeta potential of the suspension is low, then the attractive forces will dominate, resulting in flocculation of the particles. Flocculation of the particles results in the formation of a percolation network. The formed network will be viscoelastic in nature. Therefore, the network has an innate ability to sustain stress. The minimum stress that can be sustained by the formed network due to the attractive force is called static yield stress. If the applied force is greater than the attractive forces that can be sustained by the network, then the particles in the suspension start to deform, resulting in a complete or partial flow of the suspension. During the flow, the particles in the suspension interact with each other as well as with the suspending medium, thereby producing a variation in the forces. This leads to variations in the internal resistance of the suspension. Based on the variation in the internal resistance or apparent viscosity of the suspension with respect to the applied shear, the suspensions can be classified as Newtonian or non-Newtonian.

The apparent viscosity of the suspensions does not vary with respect to applied shear for Newtonian suspensions. Conversely, for non-Newtonian suspensions, the apparent viscosity of the suspension changes with respect to applied shear. On shearing a suspension, based on the characteristics and dispersion of the particles, rheological behaviour of the suspension will change. The following cases can be defined based on the equilibrium between forces: Figures 1.1 and 1.2 show the typical behaviour of different suspensions.

DOI: 10.1201/9781032635392-1

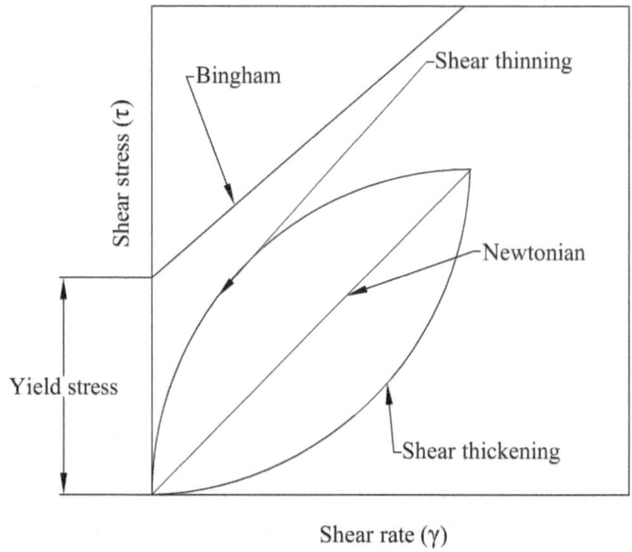

FIGURE 1.1 Different behaviours of suspension.

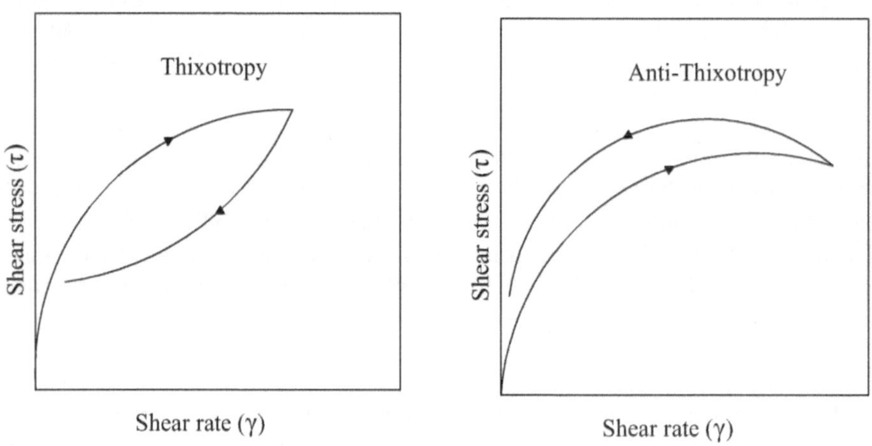

FIGURE 1.2 Thixotropy and anti-thixotropy behaviours.

1. *Attractive force > Repulsive force:* The percolation network formed will be intact. The suspension possesses an elastic network that needs energy to destruct. The particles in the suspension will be flocculated/agglomerated.
2. *Attractive force ≈ Repulsive force:* The suspension will be in a stable state. Particles will be distributed uniformly throughout the suspending medium.
3. *Attractive force < Repulsive force:* The suspension will be in the most stable state. Any further increase in repulsive force may lead to a change in internal resistance.

Shearing the suspension results in hydrodynamic forces that start to de-agglomerate or re-aggregate the particles. If the particles in the suspension are highly agglomerated, the applied shear forces will de-agglomerate the particles in the suspension. De-agglomeration of the particles decreases the apparent viscosity. Subsequently, the suspension exhibits shear-thinning behaviour. De-agglomeration of the particles in the suspension lasts until the apparent viscosity reaches a minimum value and stays at that value for a small magnitude of shear rate. A further increase in shear rate increases the apparent viscosity of the suspension due to the re-aggregation of the particles. Re-aggregation of the particles results in the formation of clusters. These clusters are formed due to hydrodynamic forces, so they are also called hydroclusters. Formation of hydroclusters increases the average size of the particles. An increase in the size of the particles increases the resistance offered by the suspension to flow. By this means, the apparent viscosity of the suspension increases. An increase in the apparent viscosity of the suspension with respect to an increase in flow rate is called shear-thickening behaviour.

Certain suspensions react or gel with time. This results in formation of a network that rigidifies as time progresses. The resulting rigidified network changes the rheological behaviour of the suspension. Consequently, the systems that rigidify have to be evaluated with respect to time in order to have a thorough and innate understanding. The ability of a system to rigidify or gel at rest, like a solid, and flow on application of a force due to the destructuration of the formed network is called thixotropy. In general, thixotropic systems are time-dependent suspensions that shear thin due to the destructuration of the formed network by an applied stress or shear and regain their structure at rest. On the contrary, certain suspensions will have time-dependent destructuration of the particles on shearing, followed by re-aggregation of the particles due to the applied stress or shear.

For any industrial process, based on the processing speed, processing rate, etc., the suspensions undergo one or a combination of the above behaviours. Therefore, it is imperative that the suspensions be tested under specific conditions in order to process them effectively and efficiently. In order to understand the change in properties exhibited by the suspension, a thorough rheological investigation needs to be carried out. The device that is used to investigate the change in behaviour of the suspensions with respect to applied shear or stress is called rheometer. In this chapter, basic terminologies, understanding about the device used for rheological measurement, and different tools will be explained in detail.

1.2 TERMINOLOGIES

Apparent viscosity: Apparent viscosity is the instantaneous viscosity. For a Newtonian material, this is similar to its viscosity. On the other hand, for a non-Newtonian material, apparent viscosity changes with applied shear rate. Apparent viscosity is the ratio between instantaneous shear stress and instantaneous shear rate.

Plastic viscosity: The resistance offered by the material to flow. This is calculated from the slope of shear stress vs. shear rate plot. Plastic viscosity remains constant for a material at a specific temperature and pressure. Plastic viscosity does not

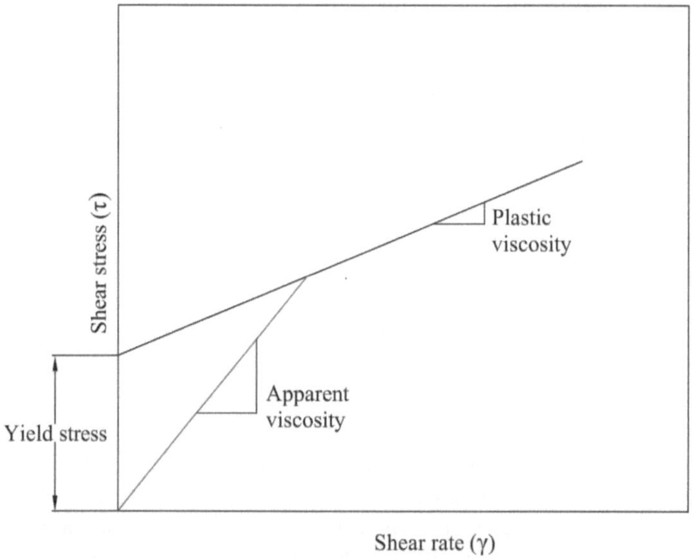

FIGURE 1.3 Yield stress of a Bingham material. Difference between apparent and plastic viscosity.

depend on the applied shear rate. The difference between apparent viscosity and plastic viscosity can be visualised in Figure 1.3.

Static yield stress: This is the magnitude of stress given to a material in order to deform the material from a solid to a liquid state. The stress given to deform the material from rest to flow is greater than the stress given to the material in order to keep it in its state of flow.

Dynamic yield stress: Once a material is deformed, it starts to flow. An adequate amount of stress is required to keep the material in its state of flow. The stress that is required to keep the material in its flowable state is called dynamic yield stress. In general, dynamic stress is lower than static yield stress.

Shear thinning: A decrease in the apparent viscosity of the suspension with respect to an increase in shear rate is called shear thinning.

Shear thickening: An increase in the apparent viscosity of the suspension with respect to an increase in shear rate is called shear thickening.

Thixotropic: A material is considered to be thixotropic in nature if the rheological properties of the material evolve with respect to time. In general, thixotropic materials are shear thinning in nature.

Anti-thixotropic or rheopectic: Anti-thixotropic materials are time-dependent, shear-thickening materials.

Shear stress: Shear stress is the ratio between the force applied and the area of the material.

Shear rate: The rate at which the material is sheared during a flow.

Viscoelastic behaviour: A combined behaviour in which the material can act as an elastic material until a certain shear stress is reached, beyond which the viscous behaviour plays a vital role.

Storage modulus or Elastic modulus: For a viscoelastic material, storage modulus is the amount of energy stored in the material. This represents the elastic portion of the material.

Loss modulus or viscous modulus: For a viscoelastic material, loss modulus is the amount of energy dissipated by a material as heat. This represents the viscous portion of the material.

Loss factor: It is the ratio between loss modulus and storage modulus. This value can be used to identify the degree of elasticity and viscosity of the material under examination.

1.3 INSTRUMENTATION

Rheometers are used in order to understand the rheological behaviour of any suspension. Based on different setups, a rheometer can be broadly classified into rotational and oscillatory rheometers. Due to advancements in the field of sensors, today's rheometers can perform both rotational and oscillatory testing. A rheometer mainly consists of a motor, a position sensor, a transducer, a closed-loop circuit, a bearing, a controller and a measuring system. Based on the position of the motor and the transducer in the rheometer, it can be classified as a single-headed or a separate-headed rheometer (Mezger 2006). A typical rheometer is shown in Figure 1.4.

1.3.1 Single-Head Rheometer

In a single-headed rheometer, the motor and the transducer are mounted side by side. Therefore, the torque or strain that can be transformed into the measuring system comes from one side, while the measuring system on the other side is stationary. For example, if a cup and a bob are used as a measuring system, then either the cup or the bob can move based on the position of the motor and the transducer.

The operation can be done in two modes. In the first mode, the shear rate, shear deformation or strain can be controlled. In this mode, shear rate is given as the input, and the resulting stress can be detected as the output. Pre-defining the required shear rate presets the required angular speed and deflection angles as input. This can be achieved by the position sensor provided in the rheometer. The output stress can be identified by the torque experienced by the motor due to the sample in the measuring system. In the second mode, the torque given by the motor is preset. As a result, the stress acting on the suspension will be defined as input. The strain deformation experienced by the suspension is taken back as output. A schematic of a single-head rheometer is shown in Figure 1.5a.

1.3.2 Separated-Head Rheometer

In a separated-head rheometer, the motor and the sensor that measure the torque are decoupled and placed on different sides. As the motor does not control the torque applied, this type of rheometer can be used only in the first mode, where the shear rate, shear deformation or strain can be given as an input parameter. In general, the motor and the position sensor are placed on the bottom portion of the measuring system.

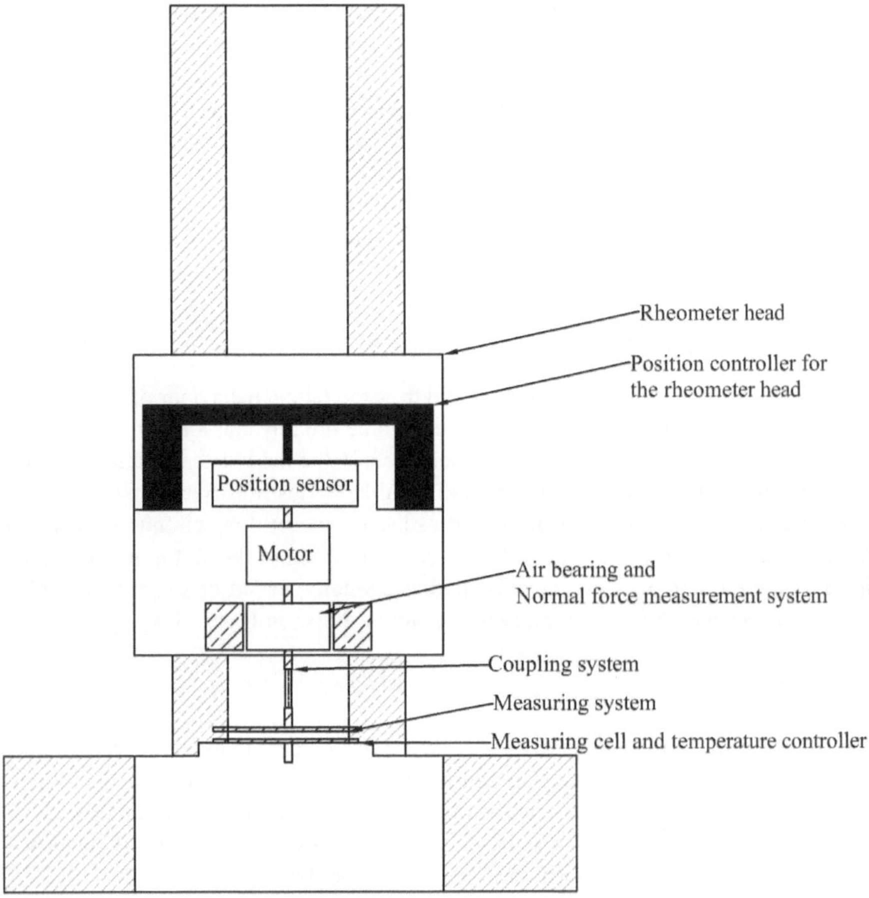

FIGURE 1.4 Single-head rheometer.

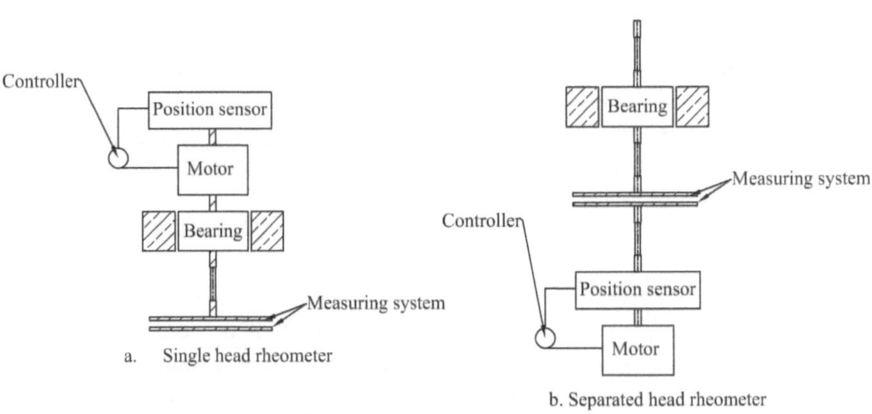

FIGURE 1.5 Rheometers with different heads.

For example, in the case of a cup-and-bob measuring system, the motor and the position sensor are placed beneath the cup, while the torque sensor is placed above the bob. In this case, the bob will be immobile. These test setups are not economical in nature because of the multiple heads in the rheometer. Figure 1.5b shows the outline of a separated-head rheometer.

1.3.3 Control Loops

1.3.3.1 Closed Control Loop for Torque

In this case, the torque is preset, and the resulting change in rotational speed is received as output. To produce the preset torque in the motor, an electronic controller directs suitable current to the motor. Successively, due to the resistance of the investigating suspension, a restoring torque will be produced. This torque will act on the motor. Meanwhile, a position sensor (an encoder or a tachogenerator) is used to measure the resulting deflection in the measuring system due to the restoring torque. This can be converted to deflection angle and rotational speed. However, other than the restoring torque from the sample under consideration, based on the bearing of the rheometer and the mass of the measuring system used, corresponding corrections need to be made for friction and inertia, respectively. This can be carried out by adjusting the current input from the electronic controller.

1.3.3.2 Closed Control Loop for Deflection Angle

Contrary to the closed control loop for torque, in this case, the angle of deflection or the rotational speed is controlled by the electronic controller. The electronic controller allows current in such a way that the torque produced by the motor is sufficient to achieve the preset angle of deflection or the rotational speed. Application of torque to the suspension results in opposing or restoring torque from the suspension to the motor. This results in regulation of the current sent to the motor by the electronic controller. Thus, the electronic controller regulates the current until the angle of deflection reaches the preset angle of deflection as input. In order to carry out the process of regulating the preset angle of deflection, a closed-loop circuit is being used that eventually decides the operating current for the required angle of deflection.

1.3.3.3 Measurement of Torque

Torque is measured either by a mechanical torque sensor or by motor power consumption. In the case of a mechanical torque sensor, a spring system is being used. When a small torque is being applied by rotation, based on the spring constant of the spring being used, force and deflection can be related. Deflection in the spring system due to the application of torque can be noted by using a mirror and high-beam light system. Based on the torque applied to the spring, deflection in the spring can be measured at a small torsion angle. The angle of deflection and spring constant can be used to calculate the resulting force or torque in the spring. Although spring systems can be used as a method to measure the angle of deflection and torque, their precision is low. Therefore, they are used only in viscometers.

In the case of modern rheometers, the current required by the motor to produce a certain torque can be measured by measuring drives. This current, in turn, can be corrected or adjusted in order to produce the required torque. Moreover, in the case of torque measurement by power consumption of the motor, any exceeding amount of torque can be produced at very low deflection. On the other hand, in the case of a spring system, a significant angle of deflection is required in order to produce the required torque.

1.3.3.4 Measurement of Deflection Angle or Rotation Speed

Whenever a torque is being applied by a motor, the change in the deflection of the motor shaft due to the restoring torque of the sample under investigation needs to be measured. This measurement is carried out either in radians or in degrees. The angular velocity of the motor shaft can be calculated by the change in the angle of deflection and the corresponding change in the time interval. In the case of viscometers, a tachogenerator that works on the principle of an inverted electrometer can be used. The working principle consists of generating current while the shaft of the motor rotates in a magnetic field. Based on the number of rotations or speed of rotation, a specific magnitude of current is produced. This current, in turn, is used to calculate the angle of deflection or the angular velocity. The tachogenerator cannot be used in the case of a static or small amplitude oscillatory shear (SAOS) test, where the measurements will be conducted at low angle deflection. Due to advancements in technology, rheometers use opto-electronic encoders. These encoders consist of a disc with stripes that are distributed evenly on the circumference. Whenever a torque is applied by the motor to the sample, based on the restoring torque to the motor, the deflection of the motor shaft is measured. This measurement is made by using a monochromatic beam of light that precisely records the start and position of the motor shaft by scanning optically. Each stripe on the disc is counted by the light pulse in an incremental measurement. This, in turn, is converted into deflection angle and angular speed.

1.3.4 Bearings

Bearings are used to prevent contact between two objects that are in relative motion. In the case of a rheometer, the motor shaft is in relative rotation with respect to a stationary cover. A ball bearing is used for this purpose in most practical applications. In a rheometer, the total torque is the sum of the torques due to the rheological behaviour of the suspension under investigation, internal friction and inertial effects due to the net mass of the rotating system. In general, internal friction and inertial effects need to be kept as low as possible. In most cases, mechanical and air bearings are used.

1.3.4.1 Mechanical Bearings

Ball bearings and roller bearings are used in low-precision rheometers and viscometers. Static and sliding friction happen due to the ongoing rotation inside the rheometer. In due course, a recalibration or change in bearing is required in order to eliminate the influence of internal friction of bearings on the test results. Therefore, a periodic check needs to be carried out at no-load conditions (without any sample) in order to nullify the errors due to the internal friction of bearings.

1.3.4.2 Air Bearings

In the case of rheometers with air bearings, the rotating shaft of the rheometer is floating on air with a continuous supply of air pressure from an air compressor. In the case of an air bearing, the torque due to internal friction of the rotating shaft is reduced to a minimum because of limited or no contact. A constant pressure source with clean, dry and filtered air is required in order to supply dust-free air to the air bearing. In order to use air bearings optimally, the following conditions need to be satisfied:

- Concentricity of the bearing should be maintained.
- The internal friction of the bearing should be minimal; i.e., the occurrence of a small scratch on the bearing surface may cause uncontrolled and unwanted rotational motion of the rotor.
- The stiffness of the bearing should be maintained at a high level in order to prevent any movement of the rotor due to axial, horizontal or lateral loads.

Other than the above parts, the temperature of the sample can be maintained by regulating the temperature of the measuring system. Variations in temperature can be controlled by different methods. Liquid bath, electric heating, heating and cooling with air or gas, heating by induction and heating by Peltier elements.

1.4 TYPES OF MEASURING SYSTEMS

Different measuring systems are used according to specific requirements. For example, for a low-viscous material, concentric cylinders are used as a measuring system. For highly viscous and viscoelastic materials, parallel plates can be used. On the other hand, for all other samples, a cone and plate can be used. In this section, different measuring systems, along with their advantages and limitations, are discussed (Mezger 2006).

1.4.1 Cup and Bob, a Concentric Cylinder or a Co-axial Measuring System

This measuring system consists of a bob (inner cylinder) and a cup (outer cylinder), as shown in Figure 1.7. In this system, the inner and outer cylinders are concentric to each other during the working position. The sample is filled into the narrow gap that is formed between the two concentric cylinders. As long as the gap between the concentric cylinders is thin, the applied torque or deformation is considered to be uniform and spread throughout the sample. If the gap between the cylinders is wide, then the applied stress will not be uniform, which results in shear-induced particle migration. Moreover, instabilities may happen due to secondary flows, turbulent flow in the case of low-viscous materials, inhomogeneous deformation, etc.

Based on the heads in the rheometer, this measuring system can be operated in two ways for rotational rheological testing. They are the Searle method, in which torque or deformation is given to the bob or the inner cylinder while the cup or the outer cylinder remains idle. The major disadvantage with this method is that at high shear rates or stress, for a sample of low viscosity, flow may change into turbulent. This leads to 'Taylor vortices' because of the centrifugal and inertial effects of the sample flow. On the other hand, in the couette method, the cup or the outer cylinder is subjected to torque or deformation while the bob or the inner cylinder remains stationary. One of the advantages of this method is that the turbulent flow at high shear rates for low-viscous samples can be minimised. Therefore, the presence of Taylor vortices can be nullified, although the major disadvantage is that the temperature control system needs to be redesigned in order to cover the entire unit, or sealant needs to be used above the gap of the cylinders in order to maintain the temperature. The use of sealant contributes to the resistance offered, along with the resistance offered by the sample under investigation that cannot be nullified.

In general, the gap between the cylinders is set as narrow as possible for the samples to experience uniform shear stress or shear rate. Applied shear stress and shear rate at any point of the gap in a cup-and-bob system can be calculated by equations 1.1 and 1.2:

$$\tau(R_n) = \frac{T}{2\pi R_n^2 L} \qquad R_i \le R_n \le R_o \tag{1.1}$$

$$\dot{\gamma}(R_n) = \frac{2R_i^2 R_o^2 \omega}{R_n^2\left(R_o^2 - R_i^2\right)} \qquad R_i \le R_n \le R_o \tag{1.2}$$

where $\tau(R_n)$ is the shear stress, T is the torque applied, L is the length of the bob, R_n is the radius of the concentric cylinder at which shear stress or shear rate need to be measured, $\dot{\gamma}(R_n)$ is the shear rate, ω is the angular velocity, and R_i and R_o are the radius of the bob and the inner surface of the cup. If the gap between the cylinders is considered to be wider, the viscosity of the sample to be investigated at the surface of the bob is given by equation 1.3:

$$\eta = \frac{\tau(R_i)}{\dot{\gamma}(R_i)} = \frac{R_o^2 - R_i^2}{4\pi L R_o^2 R_i^2}\frac{T}{\omega} \tag{1.3}$$

For a wide-gap measuring system, the applied shear stress or the shear rate will be the maximum at the bob that decreases and reaches a minimum at the inner surface of the cup. It is desired to maintain a constant shear force across the gap. Therefore, based on the requirement, narrow as well as wide-gap measuring systems are being used. As mentioned earlier, in a wide-gap measuring system, the distribution of the shear rate may not be uniform across the cross section, especially in the case of low-viscous samples. Consequently, inhomogeneous deformation may take place in viscoelastic samples. Due to this, the shear rate and the velocity gradient across the gap may not be uniform, as shown in Figures 1.6 and 1.7. In general, the ratio of the outer to the inner radius must be maintained at a certain value for the shear to be uniform across the gap. Recommendations show that the maximum allowable limit for the ratio of the outer to the inner radius is 1.0847. Similarly, typical dimensions of a cup-and-bob measuring

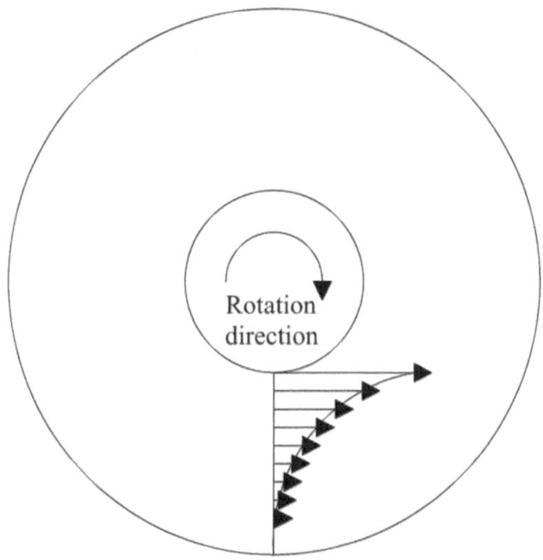

FIGURE 1.6 Velocity gradient in a cup and a bob measuring system.

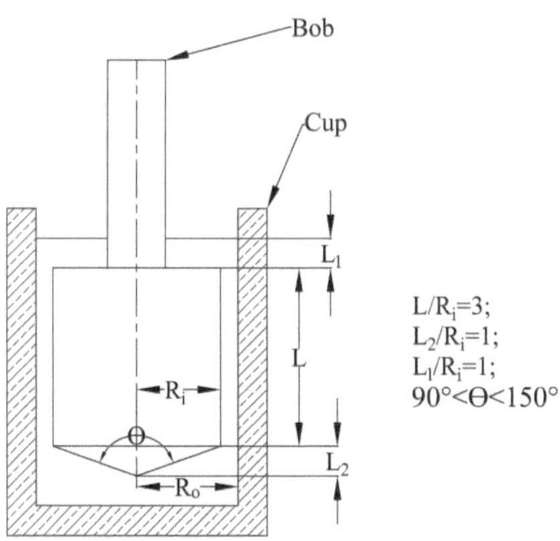

FIGURE 1.7 Cup-and-bob measuring system.

system are shown in Figure 1.7. However, because the gap between the cylinders is narrow, the distribution of shear stress and shear rate may not be uniform across the gap. Therefore, the average shear stress is equivalent to the sum of the shear stress at the inner and outer cylinders. Consequently, shear stress can be given as shown in equation 1.4:

$$\tau = \frac{\tau_i + \tau_o}{2} = \frac{\left(1+\Delta^2\right)}{2\Delta^2}\frac{T}{2\pi L R_i^2 C_E} = C_\tau T \tag{1.4}$$

where Δ is the ratio of the radius of the outer cylinder to that of the inner cylinder and C_E is a correction factor that accounts for the torque occurring at the apex of the inner cylinder, typically considered as 1.10. Eventually, each measuring system will have a constant C_τ that converts torque to a corresponding shear stress and vice versa. C_τ depends only on the dimensions of the geometry under consideration. As the radius of the bob or the inner cylinder increases, the magnitude of Δ tends to be zero. As a result, the sensitivity of the system to measuring torque at low values increases. In practice, for low-viscosity samples, a bob or inner cylinder of increased radius is useful. On the other hand, the radius of the bob should be small for highly viscous samples.

1.4.2 Advantages of Using the Cup-and-Bob Measuring System

1. Control of temperature across the sample is much easier and more efficient.
2. Slip at the surfaces can be avoided or minimised by using sandblasted or profiled cylinders.
3. The rod climbing effect does not have much effect on the sample spilling out of the cup.
4. The samples that are highly flowable in nature can be held in the cup-and-bob system, which is not possible in the case of parallel plate or cone-and-plate measuring systems.

1.4.3 Limitations of the Cup-and-Bob Measuring System

1. The quantity of samples required is high.
2. At high rotational speeds, instabilities may occur while testing low-viscous samples.

1.4.4 Cone-and-Plate Measuring System

A cone-and-plate measuring system consists of a relatively flat plate with a cone angle less than 1° and a radius of 10 to–100 mm. Typically, the cone part is set in motion, while the plate part remains stationary. Similar to a cup-and-bob system, the shear stress and the shear rate in the cup and plate system are given by equations 1.5 and 1.6, respectively:

$$\tau = \frac{3T}{2\pi R^3} = C_\tau T \tag{1.5}$$

$$\dot{\gamma} = \frac{\omega}{\tan\theta} = C_{\dot{\gamma}}\omega \tag{1.6}$$

where R is the radius of the plate, ω is the angular velocity, θ is the cone angle, C_τ and $C_{\dot{\gamma}}$ are conversion constants, and N is the number of rotations per minute. Analogous to the cup-and-bob measuring system, in a cone-and-plate measuring system as well, to increase the sensitivity at low torque values, a measuring system with

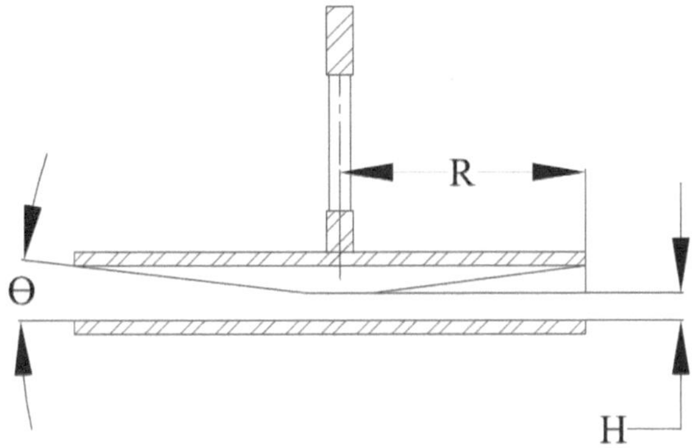

FIGURE 1.8 Cone-and-plate measuring system.

a larger radius is used. On the other hand, a measuring system with a smaller radius can be used for a highly viscous sample. From equation 1.6, it can be understood that $C_{\dot{\gamma}}$ only depends on the cone angle and is independent of the dimension of the cone used. Consequently, the shear rate will be constant across the radius of the system. The shear rate and angular velocity conversion constant $C_{\dot{\gamma}}$ is inversely proportional to the cone angle. Consequently, in order to achieve higher shear rates, a lower cone angle is preferred at constant angular velocity. Similar to the cup-and-bob system, secondary flows and flow instabilities may happen if low-viscous samples are tested at high shear rates. On the contrary, for viscoelastic samples, even at low shear rates, flow instabilities and particle migration may happen due to the elasticity of the sample under investigation. Even in some cases, the cones are truncated to have the following advantages:

1. The wear due to abrasion can be reduced. Consequently, changes in the geometrical conditions are reduced.
2. Due to the truncation in the cone, there won't be any direct contact between the cone and the plate. Therefore, parasitic friction can be avoided.

However, the maximum size of the particle that can be used during the investigation in the case of a cone-and-plate measuring system is restricted to H/5, where H is the height between the lower plate and the cone. If the necessary gap size is not provided, this may lead to direct contact between the particles and the measuring system, which may result in an increased frictional contribution to the test results. In the case of low-viscous samples, due to the applied shear, the samples may migrate off the cone-and-plate system. In order to avoid it, the sample is overfilled to a distance of 1 mm beyond the edge of the cone. A representation of a cone-and-plate measuring system is shown in Figure 1.8.

1.4.5 Advantages of Using a Cone-and-Plate Measuring System

The following are the advantages of using a cone-and-plate measuring system:

1. The applied shear rate is almost constant throughout the gap.
2. The sample required is very small.
3. Air bubbles generated during the sample preparation are excluded by setting the cone plate measuring system.

1.4.6 Limitations of Using the Cone-and-Plate Measuring System

The following are the limitations of using the cone-and-plate measuring system:

1. The maximum size of the particles that can be used is limited.
2. While testing highly viscous or viscoelastic samples, the time required for the samples to reach equilibrium is quite high. Therefore, a longer period of rest is required.
3. Flow-related inhomogeneities such as fracture on the surface, migration off the gap, solvent evaporation, turbulent flow and inertial effects may occur during the testing.
4. Based on the position of the temperature control, a temperature gradient may take place in the sample. Therefore, the temperature gradient can be minimised by using special controls that enclose the complete measuring system.
5. Profiling or sandblasting of cone plate measuring systems results in permanent changes in the dimensions of the measuring system. Therefore, it is recommended not to use sandblasting or profiling.

1.4.7 Parallel Plate Measuring System

A parallel plate measuring system consists of two plates parallel to each other. Typically, the diameter of a parallel plate system is between 20 and 50 mm. The recommended gap between parallel plates will be in the range of 0.5 to –3 mm. In general, $2R/H$ will be in the range of 10 to–50. Here, R is the radius of the plate and H is the gap between the plates. The roughness of the plate is limited to a maximum value of 0.25 μm. An increase in the gap between parallel plates may have the following effects during the investigation of different samples:

1. Secondary flow effects during the investigation of low-viscous samples.
2. Time-dependent effects in polymeric samples.
3. Edge failure in viscoelastic samples.
4. Inhomogeneous deformation in paste-like samples.

In the case of samples investigated in oscillatory shear tests, especially in SAOS, larger gap between the plates does not have a considerable effect, exclusively if the samples are investigated within their linear viscoelastic domain. Since the shear stress applied deforms the sample homogeneously. The shear stress and shear rate experienced by a parallel plate system are given by equations 1.7 and 1.8:

$$\tau = \frac{2T}{\pi R^3} = C_\tau T \tag{1.7}$$

$$\dot{\gamma} = \frac{R\omega}{H} = C_{\dot{\gamma}}\omega \tag{1.8}$$

where R is the radius of the parallel plate under consideration, and H is the gap between the plates. From the equations, it can be understood that the shear stress is inversely proportional to the radius of the plate, and the shear rate depends on both the radius and the gap between the plates. An increase in the radius of the plate increases the sensitivity at low shear stress. The shear rate is directly proportional to the radius of the plate. Therefore, the shear rate is not constant throughout the sample. It will be minimal at the centre of the plate. It increases and reaches a maximum at the edge. Consequently, the shear rate experienced by the sample will be the maximum at the edge of the plate. In addition to the radius of the plate, the shear rate is inversely proportional to the gap between the plates. An increase in the gap between the plates decreases the maximum shear rate that can be imposed by the parallel plate measuring system on the sample. On the contrary, a decrease in the gap increases the maximum shear rate that can be applied to the sample. Similar to other measuring systems, at high shear rates for low-viscous samples, secondary flow effects occur that lead to turbulence and increased flow resistance. Similarly, for high-viscous samples at low shear rates, instability in flow and shear-induced particle migration that throws off the sample from the measuring system may occur. Identifying an appropriate gap is one of the important criteria when testing different samples.

Figure 1.9 shows a parallel plate measuring system. Based on the previous recommendation, for a resin or polymer melt, the gap should be maintained between 0.5 and 1 mm. According to ASTM D4440, a gap of 1 to–3 mm is recommended for testing thermoplastic resins. For neat resins, approximately a gap of 0.5 mm is recommended by ASTM D4473 (ASTM D4440 2015; ASTM D4473 2021). In the case of dispersions and gels, the minimum gap should be at least 5 times the maximum size of the particles. However, for any gap values less than 0.3 mm, a special analysis needs to be carried out with pre-calibrated material in order to accommodate the effect of shear heating due to the internal frictional effects in narrow gaps.

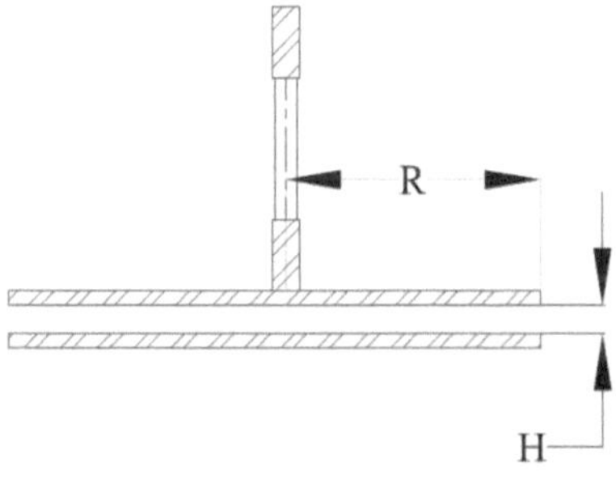

FIGURE 1.9 Parallel plate measuring system.

1.4.8 Advantages of Using a Parallel Plate System

The following are the advantages of using a parallel plate system:

1. Particles of larger size in diameter can be assessed.
2. As the set gaps are not too narrow, for samples such as polymer melts, the sample loading time is shortened.
3. When a test is performed over a wide range of temperatures, temperature gradient in the sample can be reduced by choosing a wider gap.
4. Slip can be reduced by using plates that are either profiled or sandblasted.

1.4.9 Limitations of Using a Parallel Plate System

The following are the limitations of using a parallel plate system:

1. Shear is not constant; it is zero at the centre of the plate and increases to a maximum at the edge of the plate.
2. Inhomogeneity due to flow, turbulence or migration of the sample away from the plate may happen near the rotating plate.

1.5 FACTORS INFLUENCING RHEOLOGICAL PARAMETERS

In general, numerous parameters affect the rheological behaviour of suspensions. Influencing parameters can be broadly classified into three heads. They are the physical, chemical and surface characteristics of the material. Some of the parameters are solid volume fraction (SVF) or water-to-binder (w/b) ratio, particle size distribution, specific surface area (SSA), reactivity, particle shape, viscosity of the suspending medium, temperature, etc.

1.5.1 Viscosity of Suspending Medium

Generally, suspending medium will be Newtonian in nature. An increase in the viscosity of the suspending medium increases the overall viscosity of the suspension. As a general rule, variations in the viscosity of the suspending medium do not affect the yield stress of the suspensions. If a solid element is immersed in the suspending fluid, the movement of the element in the suspending medium undergoes resistance due to the hydrodynamic force induced by the flow of the surrounding medium. In the case of a Newtonian suspending medium, the resistance is proportional to the shape, fluid viscosity, characteristic length and relative velocity between the solid and the suspending medium. Moreover, for a suspension with a low SVF (less than 4%), the immersed solids are far away from each other. Therefore, the viscosity of the suspension will be almost equivalent to the viscosity of the suspending medium. On the other hand, an increase in the SVF increases the viscosity of the suspension.

1.5.2 Solid Volume Fraction

An increase in the SVF of the particles in the suspension increases the viscosity of the suspension. For example, consider a suspension of thickness 'T' sheared between two plates. The bottom plate is fixed, while the top plate can move with a certain velocity 'V'. Consider the thickness of the particles as 'Δ'. Now, the shear stress experienced by the suspension of thickness 'T' is given by $\sigma = \mu_0 \dot{\gamma}_{\text{apparent}}$. Therefore, the apparent shear rate is $\dot{\gamma}_{\text{apparent}} = V/T$. Due to the inclusion of the particles of thickness 'Δ' in the suspending medium, the effective shear rate increases. This is given by $\dot{\gamma}_{\text{effective}} = V/T - \Delta$. Based on the above changes in apparent and effective viscosity, the shear stress of the suspension for a suspending medium of thickness T and with the inclusion of thickness $T - \Delta$ is given by $\sigma = \mu_0 \dot{\gamma}_{\text{apparent}} = \mu_0 \dot{\gamma}_{\text{effective}}$. Therefore, the change in the viscosity of the suspension due to the inclusion of the particles of thickness Δ is given by $\mu_0 = \mu_0 \dot{\gamma}_{\text{effective}} / \dot{\gamma}_{\text{apparent}} = \mu_0 \left\{ 1/1 - \Delta/T \right\}$. As the thickness of the particles in the suspension increases and reaches T, then $1/1 - \Delta/T$ becomes infinite. Eventually, the viscosity of the suspension reaches infinite. With respect to increase in SVF of the particles in the suspension, the viscosity of the suspension increases (Coussot 2005).

Moreover, an increase in the SVF of the particles in the suspension decreases the available suspending medium to separate two or more particles. In due course, the particles start to interact with each other directly by colliding or rubbing against each other. This results in increased apparent viscosity of the suspension at high shear rates, i.e., the suspensions with high SVF undergo an earlier onset of shear thickening as well as increased intensity of shear thickening (Moorthi et al. 2020).

1.5.3 Size of the Particle and Rheology

The size of the particle is predominantly one of the most important factors that affect the rheological behaviour of suspensions. For a mono-disperse suspension (approximately the particles are of single size), if the size of the particle is bigger, the apparent viscosity of the suspension will be higher. On the other hand, the smaller the particles, the lower the viscosity of the suspension. An increase in size of the particle increases Δ. Apparently, the viscosity of the suspension reaches infinity.

1.5.4 Specific Surface Area (SSA)

A change in SSA profoundly affects the rheological behaviour of any suspension. For example, consider two suspensions with similar SVFs, one with a larger particle size and the other with a smaller particle size. The SSA of the suspension with a smaller particle size at a similar SVF will obviously be higher, when compared with the other suspension with bigger particles at a similar SVF. The suspension with smaller particles will have higher resistance to applied shear, since the particles with higher SSA require more suspending medium to lubricate all the particles. As a result, the suspension with smaller particles at similar SVF will have relatively higher collisions and contact with each other, which obviously increases the viscosity of the suspension. Moreover, in the case of colloidal suspensions, the higher the SSA, the higher the inter-particle attractive forces. Higher inter-particle attractive forces result in large, agglomerated particles, so the yield stress of the suspension will also be higher. Whenever the shear force increases, the particles in the suspension start to de-agglomerate due to the hydrodynamic forces thus, resulting in shear-thinning behaviour.

1.5.5 Reactivity

Most of the particles used in industrial processing are reactive in nature. With respect to time, they undergo chemical as well as physical changes. Physical and chemical changes that take place in the suspension transform the rheological behaviour of the suspension. The reactivity of the particles in the suspension determines the increase or decrease in the yield stress, viscosity, thixotropic nature, etc. of the suspension. Especially in the case of cementitious suspensions only with ordinary Portland cement (OPC), the reactivity of the particles will be higher. Therefore, with respect to time, formed hydration products increase the ion–ion correlation, thereby rigidifying the formed network. A formed network increases the strength. Consequently, the stress required to destruct the structure increases, resulting in increased yield stress in the suspension. In contrast, if OPC in the cementitious suspensions is replaced by less reactive materials such as slag or fly ash (FA), with respect to time, the rigidification process will be slow. The increase in the yield stress of the suspension will also be slow with respect to time.

A suspension with high yield stress results in de-agglomeration of the particles with respect to applied shear, resulting in shear-thinning behaviour. A further increase in applied shear results in the formation of hydroclusters. This phenomenon is responsible for the onset of shear-thickening behaviour. On the contrary, for a suspension with low yield stress, the internal resistance will be low. As a result, the shear-thickening behaviour of the suspension onsets at early shear rates (Moorthi et al. 2020). In the case of the thixotropic nature of the suspensions, the suspension with high reactivity will possess a high thixotropic nature. An increase in time increases the number of bonds formed per unit volume resulting in higher thixotropic nature of the suspension with highly reactive particles.

1.5.6 Water-to-Binder Ratio

An increase or decrease in the w/b ratio modifies the rheological behaviour of cementitious suspensions. For example, an OPC-based suspension with a low w/b ratio will have less water to disperse the particles uniformly. Consequently, their yield stress will be high as the particles will be in a highly agglomerated state. Moreover, water in the system will be less, so the reactivity of the system will be less with respect to time. As a result, there won't be any abrupt increase in yield stress in the system with respect to time. On the contrary, at high w/b ratios, the system will have enough water to disperse the particles in the suspension. The particles will be in a uniformly dispersed state, which leads to low yield stress in the system. Besides, due to the higher water content in the system, OPC in the suspension can hydrate easily. Hence, with respect to time, the yield stress increases rapidly.

In systems with a low w/b ratio, the amount of water available will be less in order to cover all the particles in the system. Consequently, the probability that a particle can interact directly with a neighbouring particle will be quite high. Therefore, the internal resistance offered by the particles in response to shear force will be higher. Additionally, with respect to an increase in shear force, the shear-thickening behaviour of the suspension can be evident. Alternatively, at high w/b ratios, the particles in the system will be well dispersed. Subjecting the suspension with a high w/b ratio to shear force results in an early onset of shear thickening and a higher intensity of shear thickening.

1.5.7 Shape of the Particles and Rheology

Similar to the size of the particle, the shape of the particles also has a major effect on the rheological behaviour of suspensions. Particles with a spherical shape enhance the flow of the suspension. Since they can slide over each other, particles with a spherical shape will have the least resistance to shear. An increase in irregularity in the surface of the particle increases interlocking between the particles; as a result, the yield stress and the apparent viscosity of the suspension increase. In the case of suspensions with fibres, in a shear flow, the fibre tends to orient in the direction of the shear flow. This tends to reduce the resistance offered by the particles in the suspension to flow. In general, addition of fibres results in shear thinning of the suspension.

1.5.8 Effect of Addition of Polymers

Polymers are added to the suspensions in order to provide repulsive forces to the particles. Thereby, the yield stress and the apparent viscosity of the suspension decrease. Specifically, in the case of cemetitious suspensions, lignosulfonates (LS), sulfonated naphthalene formaldehyde (SNF), sulfonated melamine formaldehyde (SMF) and poly carboxylates (PCs) have been used in order to increase the dispersion of the particles in the suspension. The mentioned polymers are called plasticisers or superplasticisers (SPs). In the case of cementitious suspension, these polymers act as a source of electrostatic repulsion or steric hindrance. LS, SNF and SMF act based on electrostatic repulsion, while PCs act based on both electrostatic repulsion and steric hindrance. Obviously, at similar dosages, PCs perform better than LS, SNF and SMF. Therefore, PCs effect on reducing yield stress as well as apparent viscosity is tremendous, although their usage reduces the onset of shear-thickening behaviour at early shear rates and increases the intensity of shear thickening. Besides, in suspensions with low w/b ratios, high-molecular-weight non-ionic polymers such as poly-ethylene glycols (PEGs) are used in order to delay the onset of shear thickening and reduce the intensity of shear-thickening regime change from continuous to dis-continuous.

1.6 EXPERIMENTAL PROCEDURES

Interpretation of experimental results is not easy in practice for rheological experiments, although the rheological theories provide relationships that are up-front in connecting the functions of materials to their macroscopic characteristics. In particular, typical behaviours such as yield stress, viscoelasticity and thixotropy are not directly derived from data. Since they are highly connected with momentary behaviour changes, to determine their rheological properties, it is essential to develop explicit investigational approaches. Even from the set of data obtained from dedicated experimental procedures, different supplementary effects may occur. Supplementary effects include but not limited to flow heterogeneities due to geometrical variation of the measuring system, heterogeneities in material due to migration, non-ideal boundary conditions such as edge deformation, wall slip, etc. All the above effects need to be considered in interpretation of rheological data. Subsequently, the above effects diverge from the ultimate flow behaviour of any material (Coussot 2005).

1.6.1 Sample Preparation: A Generic Paradigm

The state of the sample in its visibly well-known stage is crucial for any rheological test. A clearly identified state is necessary in order to get reproducible rheological data for identifying changes in behaviour. Whenever a sample is being prepared based on the experiment, substantial changes will occur to the internal structure of the sample due to the uncontrolled history of deformation. This may lead to significantly non-reproducible rheological data. Consequently, it is advisable to carry out a rapid shear and rest the sample for a short period of time. Depending on the pre-shear intensity and time of application, the internal structure of the sample may reach a steady state. Choosing the time of rest after the rapid pre-shear can have various significant roles. The applied pre-shear can cause substantial inertial force in the sample. Resting time may be used to circumvent the undesirable effect due to inertia caused by pre-shear. In other cases, where the sample is viscoelastic, the time of rest is decided by the sample to regain its solid nature. On the other hand, for a thixotropic sample, the time of rest plays a substantial role that signifies a different restructuring state of the sample.

1.6.2 Determination of the Solid Behaviour of the Sample: Viscoelasticity

Suspensions that are viscoelastic in nature remain solid until a certain stress value. If a stress of magnitude greater than the critical value is applied, then the suspension starts to flow. In order to identify the magnitude of stress at which the suspension starts to flow, different test procedures can be used. During the sample preparation phase and pre-shearing, the samples undergo deformation that results in transition of their regime from a solid to a liquid state. To determine the solid nature of the sample, it must be in its solid state. Consequently, after the pre-shear, the samples are left undisturbed, in order to develop their internal structure. After this, a creep or a dynamic test can be performed in order to understand the solid nature of the sample.

1.6.3 Creep Test: Determination of Static Yield Stress

The sample is subjected to a constant stress (τ) in a creep test. The corresponding strain obtained with respect to time is considered to understand the solid behaviour of the sample under investigation. Based on the applied value of the stress and the yield stress (τ_y) of the sample, two different kinds of curves can be observed. If the relative stress applied is less than the yield stress of the sample ($\tau < \tau_y$), then the sample will be in its solid regime. The initial slop between strain and time graphs will be concave in nature at the start of the experiment. As time progresses, the applied shear is not enough to deform the sample enough to transform it from a solid to a liquid state. Consequently, the curve flattens and becomes an asymptote with respect to time. However, if the applied stress is greater than the yield stress ($\tau > \tau_y$), the sample starts to deform, turning into its liquid regime. In this case, the initial part of the strain vs. time curve exhibits a similar trend to that of the curve at low stress, although with the passage of time, the curve shows a linear increase in strain after it has reached a point of inflection. A model that represents the solid regime of a viscoelastic sample in its simple form shows that the stress applied is the sum of the viscous and elastic components. Equation 1.9 shows the Kelvin–Voigt model in a simple shear flow:

$$\tau = G\gamma + \mu\dot{\gamma} \quad \gamma < \gamma_c \tag{1.9}$$

where τ is the total stress, G is the elastic modulus, μ is the viscosity, γ is the strain applied, γ_c is the critical macroscopic deformation and $\dot{\gamma}$ is the shear rate. In the context of a viscoelastic material, it can be presumed that under any deformation, that is, within the critical macroscopic deformation, the particles in the material will have the lowest potential energy. Once the force that displaced the particle from its lowest potential energy to a higher state is released, the particle will come back to its lowest potential energy. During the applied force, the particles are displaced, although their displacement is linear. Moreover, along with the elastic behaviour due to the infinitesimally small displacement, there will be viscous dissipation due to the relative motion of the particles (hydrodynamic interaction).

Consequently, due to the applied deformation, along with the solid nature of the material, a viscous nature also persists. Under the critical deformation range, the material will remain a viscoelastic solid. Once the applied deformation is greater than the critical macroscopic deformation of the material, the particles are permanently displaced from the lowest potential energy to a higher potential energy. This results in irreversible deformation of the material; therefore, the particles cannot be brought back to their initial positions permanently. As a result, only the liquid nature of the material will be dominant (Figure 1.10). Beyond the critical deformation range, the material will remain a viscoelastic liquid, although it is not clear how the elastic nature of the material contributes to the liquid regime. Meanwhile, the viscoelastic, solid nature of the material can be regained by allowing it to rest for a sufficient time. The average stress that corresponds to extracting the particles from their average minimum potential is considered to be the yield stress of the material. For the yielding criteria of a viscoelastic material, equation 1.10 remains true:

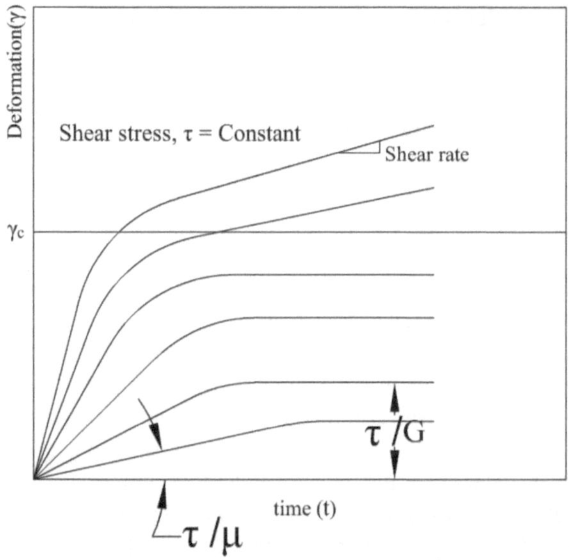

FIGURE 1.10 Creep test for increasing values of shear stress applied, γ_c is the critical deformation, below which the elastic regime is dominant above which the liquid regime is dominant (non-thixotropic material).

$$\tau < \tau_c \leftrightarrow \dot{\gamma} = 0 \tag{1.10}$$

where τ is the applied stress, τ_c is the critical stress at which the material yields and $\dot{\gamma}$ is the shear rate. From equation 1.10, it can be realised that for a stress less than the yield stress of the material, there won't be a steady and homogeneous state of flow. Moreover, under such circumstances, the material will deform locally in discrete places, forming a shear band. When applying high shear rates, the applied force will maintain the particles in flow by continuously breaking the initial network. Under such high deformation, the following two main interactions take place:

- In order for the network to reform, a definite time interval needs to be provided. Nonetheless, due to the continuous high shear, there won't be enough time for the particles to reach their lowest potential.
- Due to the relative motion of the particles, there will be a continuous dissipation of energy due to hydrodynamic interaction.

Consequently, the materials will be in the liquid regime, where they are referred to as viscoplastic fluids. As shown in equation 1.11, the applied stress will be a function of the critical stress and the deformation rate:

$$\tau = \tau_c + f(\dot{\gamma}) \quad \tau > \tau_c \tag{1.11}$$

For a material that has a linear dependency on the deformation rate, $f(\dot{\gamma})$ is given as $\mu_P\dot{\gamma}$. The model corresponds to Bingham. Similarly, for certain other materials, the dependency may not be linear with respect to the deformation rate. In such cases, $f(\dot{\gamma})$ is given as $\eta\dot{\gamma}^n$. The model corresponds to Herschel–Bulkley (HB). Here μ_P is the plastic viscosity, η is the flow index, $\dot{\gamma}$ is the deformation rate and n is the power index. Based on necessity, other simple equations were also used that can establish a relationship among the stress applied, yield stress and the deformation rate.

1.6.4 Dynamic Test: Determination of Rheological Parameters

In the case of using a dynamic test to determine the rheological properties of a viscoelastic sample, a periodic oscillatory strain is given to the sample in the form $\gamma = \gamma_0 \sin \omega t$. Here, γ is the strain applied, γ_0 is the amplitude of the strain applied, ω is the frequency and t is the time period. Under a strain value less than the critical strain of the sample $\gamma_0 < \gamma_c$, the viscoelastic solid behaviour of the material can be examined. In this case, due to the applied strain and elastic modulus, the viscous nature of the material can also be determined. Subjecting a viscoelastic sample to a periodic oscillatory strain results in different phases of stress vs. time curves with respect to strain vs. time, as shown in Figure 1.11. In Figure 1.11a, the stress and

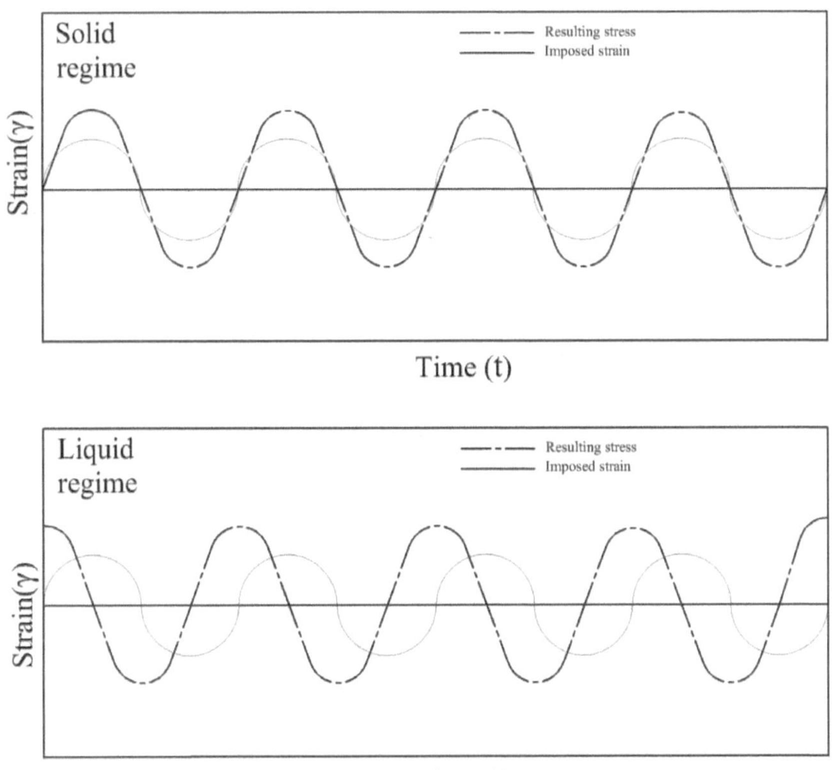

FIGURE 1.11 Dynamic test for a non-thixotropic material showing solid (stress and strain are in phase) and liquid regimes (stress and strain are out of phase).

strain plots are in phase with each other. In contrast, Figure 1.11b shows that the stress and strain plots are out of phase with each other. Stress and strain plots that are in phase with each other can be used to assess the elastic nature, while those that are out of phase can be used to assess the viscous nature of the viscoelastic sample under examination. The resulting stress can be given as shown in equation 1.12. Equation 1.12 can be rewritten as shown in equation 1.13:

$$\tau = G\gamma_0 \sin \omega t + \mu \gamma_0 \omega \cos \omega t \tag{1.12}$$

$$\tau = \tau_0 \sin\left(\omega t + \delta\right) \tag{1.13}$$

According to trigonometric laws, equation 1.13 can be further modified, as shown in equation 1.14:

$$\tau = \tau_0 \sin\left(\omega t\right) \cos(\delta) + \tau_0 \cos\left(\omega t\right) \sin(\delta) \tag{1.14}$$

Therefore, equations 1.12 and 1.14, with the following simplifications, can be arrived at as shown in equations 1.15–1.20:

$$\tau_0 \cos(\delta) = G\gamma_0 \tag{1.15}$$

$$\tau_0 \sin(\delta) = \mu \gamma_0 \omega \tag{1.16}$$

$$\tau_0 = \gamma_0 \sqrt{G^2 + \mu^2 \omega^2} \tag{1.17}$$

$$\tan\left(\delta\right) = \frac{\mu \omega}{G} = \frac{G''}{G'} \tag{1.18}$$

$$G' = \frac{\tau_0}{\gamma_0} \cos(\delta) \tag{1.19}$$

$$G'' = \mu \omega = \frac{\tau_0}{\gamma_0} \mathrm{Sin}(\delta) \tag{1.20}$$

Equations 1.19 and 1.20 can be used to determine the elastic (storage) modulus and viscous (loss) modulus of a viscoelastic sample in its solid regime. The above equations can be used for any kind of viscoelastic material.

Similar to strain oscillations, stress oscillations can also be used to obtain strain in order to determine the storage and loss moduli of the sample. For most viscoelastic materials, the solid and loss moduli remain in a simple pattern until the applied stress or strain amplitude is less than the critical deformation of the material. As a result, the storage and the loss modulus remain independent of the applied stress or strain. In the case of some other materials, the stress and strain may not remain in a simple pattern. For those materials, two tests are carried out in order to identify the dependency of the material's behaviour with respect to frequency and deformation.

- In the first case, the frequency is kept constant, while the deformation or stress applied is varied.
- In the second case, the deformation or stress applied is kept constant, while the frequency is varied.

The variation in the storage and loss moduli can be used as a basis in order to characterise the material's behaviour.

1.6.5 Stress or Shear Rate Ramp: Determination of Yield Stress

Shear stress or shear rate ramps can be used as a useful method to determine the rheological behaviour. Based on the requirement, after a pre-shear or rest time, the sample is sheared by applying a stress ramp or shear rate ramp. The critical stress at which the sample start to flow can be identified as the yield stress. Before the sample deforms and starts to flow, the shear stress gets localised. Consequently, the deformation saturates rapidly and the shear rate remains zero. Beyond the yield stress, the shear rate starts to increase rapidly. More specifically, the stress at which the sample starts to flow is not necessarily the stress at which a complete deformation of the sample can be envisaged. Therefore, in a shear stress or shear rate ramp test, an explicit transition from solid to liquid cannot be identified. Figure 1.12 shows a typical flow curve for a viscoelastic material.

1.6.5.1 Type and Magnitude of Shear Stress or Shear Rate

The magnitude and type of shear stress or shear rate given to the sample highly depend upon the practical requirement (Figure 1.13). A linear increase or decrease in the shear stress can be used for practical purposes in which the flow rate or pressure of the system increases and decreases linearly. On the other hand, constant stress or shear values for a certain period of time can be given. For example, if a suspension is

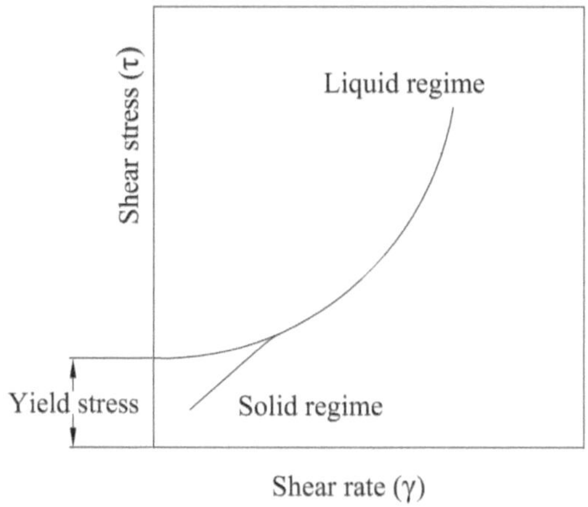

FIGURE 1.12 Shear stress vs. shear rate plot by increasing and decreasing shear stress or shear rate ramp for a non-thixotropic material.

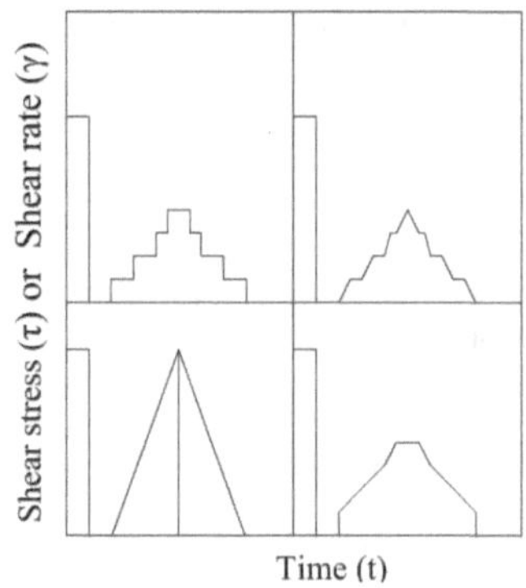

FIGURE 1.13 Different shear profiles.

processed at a rate of 30 m^3/hour in a circular pipe of 120 mm in diameter, then the velocity at which the suspension is processed is 0.74 m/s. The shear rate experienced by the sample will be the maximum at the wall of the pipe. Consequently, for a suspension without yield stress or shear thinning in nature, equation 1.21 can be used to compute the maximum shear rate that can be experienced:

$$\dot{\gamma}_{\text{maximum}} = \frac{Q}{\pi R^3}\left\{3 + \frac{1}{N}\right\} \tag{1.21}$$

where $\dot{\gamma}_{\text{maximum}}$ is the maximum shear rate, Q is the flow rate in m^3/s, R is the radius of the circular pipe in m and N is the power index of the suspension. If the suspension is shear thinning in nature, then the value of N will be less than one. The maximum value of one can be assumed, and the maximum shear rate can be determined by equation 1.21.

Assuming all the variables, a maximum shear rate of 49 1/s can be determined. This value can be used as the maximum value, and different shear profiles can be made. For a linear up-and-down profile, the minimum shear rate of 0 1/s is given and the maximum shear rate of 50 1/s is taken. The data obtained from the down-profile is considered for the analysis. Since the data in the down curve will have good stability compared to the data in the up curve, similarly, if a process involves applying a constant pressure or constant outflow, then a constant shear rate is applied in order to understand the behaviour of the suspension. In the case of suspensions with yield stress, other equations are used to theoretically estimate the maximum shear rate that can be experienced by the suspensions. Based on the rate of shear and the magnitude of maximum shear in a process, the shear profile can be modified in order to understand the behaviour of the samples in a specific condition. This may aid in understanding the behaviour of the sample in a particular process, enhancing the efficacy of the process.

1.6.6 Dynamic Test: Generic Behaviour of Viscoelastic Material

When a sinusoidal strain is applied to the sample, based on the elastic and viscous nature of the material under examination, the dynamic test can be used to infer the

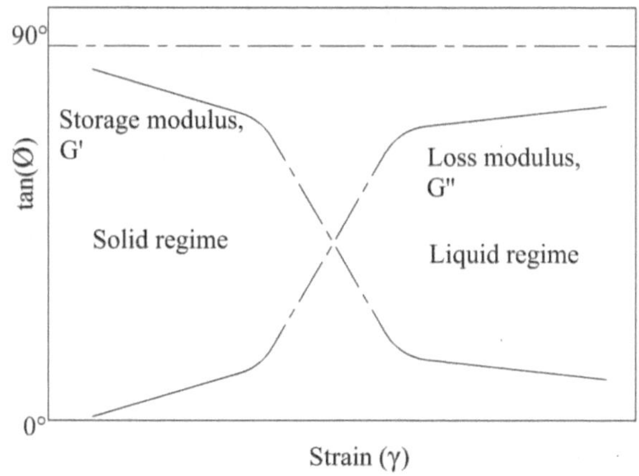

FIGURE 1.14 Variation of storage and loss modulus as a function of strain for a non-thixotropic material that corresponds to the solid–liquid transition.

relative characteristics of the sample. This is achieved by understanding the storage and loss moduli of the material (Figure 1.14).

If the sample deforms and transforms from a solid to a liquid state, then the storage modulus reaches a minimum, while the loss modulus increases (Coussot 2005). In principle, for a deformation less than critical deformation ($\gamma < \gamma_c$), the material being examined remains a solid. The real value of the stress can be generalised as given in equation 1.22:

$$\tau = \frac{G(|\gamma|)|\sin \omega t|}{\sin \omega t} + \frac{H(|\dot{\gamma}|)|\cos \omega t|}{\cos \omega t} \tag{1.22}$$

From equation 1.22, it can be understood that, at low frequencies, elastic behaviour prevails, whereas viscous behaviour prevails at sufficiently high frequencies. Consequently, storage modulus will predominate and remain at low frequencies; with respect to increases in frequency, loss modulus increases. In this case, if the deformation is greater than the critical deformation ($\gamma > \gamma_c$), then the elastic nature of the sample will be almost negligible, which drops the storage modulus to a minimum or almost negligible. On the other hand, the transition of the sample into its liquid regime results in the stress being completely dependent on the shear deformation. For a sample deformed by a controlled deformation, the stress will be as given in equation 1.23:

$$\tau = \frac{|\cos \omega t|}{\cos \omega t}\left[\tau_c + F|\gamma_0 \omega \cos(\omega t)|\right] \tag{1.23}$$

1.6.7 Experiments to Understand Thixotropic Nature

Thixotropy is a reversible phenomenon in which the suspension at rest starts to form a network. This results in an increase in the apparent viscosity of the suspension. On applying shear, the suspension starts to destructure. Consequently, the apparent viscosity of the suspension decreases. This reversible nature of the suspension between structuring at rest and de-structuring due to shear is called thixotropy. At the microscopic level, when a suspension is at rest, the particles in the suspension have the lowest potential energy. Therefore, they are surrounded by other particles in equilibrium. The colloidal forces and the Brownian forces keep the particles in their positions. As time progresses, the particle in the suspension goes to a lower potential energy due to the interconnection between the neighbouring particles. The significance of the above change is that the strength of the network between particles increased. On applying shear force to the suspension, the network between the particles starts to dislocate. Thereby, the particles in the suspension start to move from their ground level to a higher potential. Consequently, the displaced particles possess the energy to move from one location to another. Overall, the time of rest of the suspension affects the nature of structural evolution. The study of the thixotropic nature of suspension involves subjecting the suspension with respect to time and understating their behaviour with respect to time. For a thixotropic material during a creep test at the solid–liquid transition, the yield stress of the material increases with time. The increase in yield stress of the material at different times can be examined by creep test, although the time duration for the application of creep test should be less than the time taken by the structure of the material or the suspension to form. In general, the time involved in the examination of a sample by creep test is higher. Consequently, the condition of attaining the test time of the sample within the time of re-structuration of the sample may not be possible. Therefore, creep tests cannot be used as an ideal method to evaluate the thixotropic nature of the material (Coussot 2005).

On the other hand, dynamic tests such as SAOS can be used to understand the evolution of storage as well as the loss modulus of material. Provided that the applied oscillatory strain is less than the critical strain of the sample under investigation, the characteristic time of the applied strain in the case of oscillatory strain is short. Therefore, after the prescribed time interval, an oscillatory strain of magnitude less than the critical strain can be given in order to understand the thixotropic nature of the material.

For a material in its liquid regime, a continuous shear stress or shear rate can be applied, and the resulting shear rate or shear stress can be examined with respect to time (Figure 1.15). The applied shear stress or shear rate should be in the liquid regime. In order to achieve that the material is in a liquid regime, it is sheared at a higher shear rate than its critical shear rate. Now that the structure of the material is in its sheared state, this can be understood by the constant magnitude of shear stress on the material while shearing. When the shear rate is increased to a higher value and kept constant, the corresponding shear stress also increases suddenly and then decreases to a constant value. The sudden increase in shear stress due to the increase in shear rate is due to the fact that at the previous shear rate (i.e., at a lower magnitude), the particles in the material were not in a completely destructured state. Consequently, an increase in shear rate destructures all the particles in the suspension and allows it to reach an alternative equilibrium state. The sudden increase in shear stress of the material due to the increase in shear rate is called stress overshoot. Commonly, the stress overshoot increases with respect to the time of rest. On the

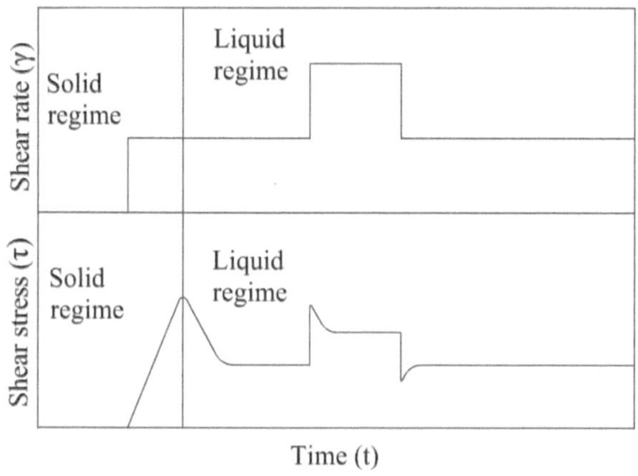

FIGURE 1.15 Shear stress vs. shear rate variation with respect to time for a thixotropic material. (Shear rate is imposed as input.)

other hand, if the shear rate is decreased suddenly and maintained constant at a certain level, the corresponding shear stress decreases suddenly, increases and remains constant. A sudden decrease in shear stress is the outcome of a more de-structured state of the particles due to the high shear rate before decreasing the shear rate to lower values. This can be seen in Figure 1.15.

Other than the above methods, a linear increase and decrease in shear rate or shear stress at specific time intervals can also be used as a test method to understand the thixotropic nature of the material. A plot between shear stress and shear rate gives a loop, also called a '*hysteresis loop*'. For a thixotropic material, the increasing and decreasing stress curves do not overlap each other. Although this method can give us a verbal idea about thixotropy, the effect of time alone cannot be separated. Therefore, using this method requires caution.

1.7 ERRORS IN RHEOLOGICAL MEASUREMENT AND HOW TO RECTIFY

Like any other sophisticated measurement method, examination of a sample in rheological measurement also has its share of contributions originating from errors. This section explains some of the common errors during a rheological experiment and makes suggestions to rectify them.

1.7.1 SHEAR-INDUCED PARTICLE MIGRATION

When a sample is sheared in a measuring system, based on the shear stress distribution, the particles in the system start migrating from one portion to another. In general, the particles start to migrate from a high shear portion to a low shear portion. Therefore, based on the movement and geometry of the rotating system, the shear-induced particle migration also varies. In the case of suspensions that are multi-disperse in nature (particles of different sizes), particles of larger sizes start to migrate quickly. This results in the formation of a thin depleted layer (a layer with no particles or particles less than the average concentration of the rest of the portion) nearby or right next to the portion where the shear stress is the maximum. This layer is often referred to as the slip layer or wall depletion layer in rheological investigations. Formation of a thin, depleted layer results in slip at the interface. Slip at the interface can be identified by a sudden increase in velocity from zero to maximum (Figure 1.16).

A decrease in the concentration of the particles in this layer decreases the apparent viscosity of this layer compared to the rest of the bulk. Consequently, the slip layer possesses an apparent shear rate that is higher than the rest of the portion away from it at a constant shear stress applied. Slip during a rheological experiment underestimates the apparent viscosity of the bulk due to the low shear rate being imposed. Slip during a rheological experiment can be avoided by providing a roughened surface or a protruding rib at the interface where there is a probability for the formation of the slip layer. In general, the thickness of the roughness or the protruding rib can be determined based on the particle size distribution. For a multi-disperse suspension, the roughness is approximately equivalent to the maximum size of the particle in the suspension. On the other hand, for a suspension with uniform particles, the roughness should be several times that of the particle size. For paste-like material, instead of the roughened surface of the cylinder, a vane can be used. Although the use of protruding ribs can eliminate wall slip effects at high shear stress, at low shear stress, it results in a localised fracture of the sample.

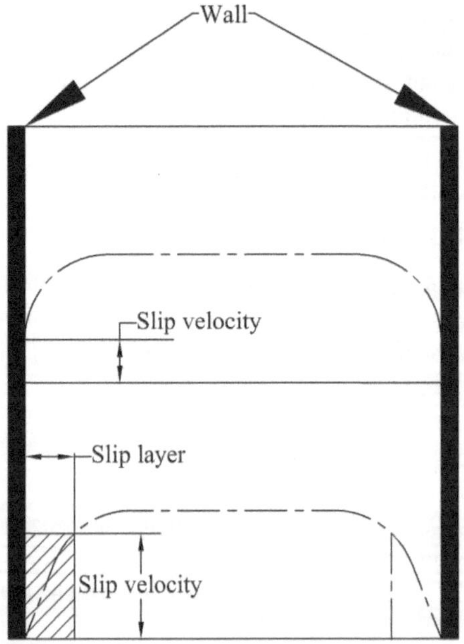

FIGURE 1.16 Wall slip.

1.7.2 Effects of Inertia and Turbulence

For any suspension, an increase in applied shear force increases the momentum of the particles. This results in the origin of inertial effects. When the momentum of particles increases, the kinetic energy of the particles in the suspension exceeds viscous dissipation. This results in flow instability, referred to as turbulence. The inertial effect of the particles in the suspension affects the rheological examination of the suspension. Consequently, it is important to understand the effects of inertia and turbulence. If the ratio of the inertial force to the viscous force is less than one, then the effect of inertia can be considered negligible during the transient character of the flow. The dimensionless number can be calculated as shown in equation 1.24:

$$Re = \frac{\rho \dot{\gamma} d^2}{t\tau_c} \tag{1.24}$$

where *Re* is the ratio between inertial force and viscous force, ρ is the density, $\dot{\gamma}$ is the shear rate, d is the thickness of the material being sheared, t is the characteristic time and τ_c is the critical stress at which the suspension deforms. In the case of geometries that are rotating in nature, the value of equation 1.25 should be less than one:

$$Re = \frac{\rho \dot{\gamma}^2 d^3}{R\tau_c} \tag{1.25}$$

where R is the radial distance in the sample. For turbulence to occur in yield stress suspensions, the ratio of kinetic to viscous energies should be much larger than one, as shown in equation 1.26:

$$Re = \frac{\rho v^2}{\tau_c} \tag{1.26}$$

In this case, τ_c is the yield stress of the suspension under consideration. Nonetheless, in the above cases, the suspension may not be experiencing high shear deformation. In those cases, equation 1.27 is sufficient to understand the effect of turbulence:

$$Re = \frac{\rho v d}{\mu} \tag{1.27}$$

where μ is the viscosity of the sample that should be replaced by apparent viscosity at high shear rates. If *Re* according to equations 1.26 and 1.27 are more than one, then turbulence will occur. The effect of inertia or turbulence on the rheological examination can be reduced by choosing an appropriate measuring system and the position of its measurement.

1.7.3 Drying and Phase Separation during Testing

In general, any long-term testing results in the drying of suspensions due to the evaporation of the suspending medium. Drying results in an increase in the SVF of the particles in the suspension. Therefore, the rheological measurements are not with respect to the initial SVF of the suspension. In order to reduce the effect of drying on the rheological measurement, certain countermeasures can be taken.

- Larger is the surface of the sample exposed to the atmosphere, higher will be the evaporation of the suspending medium. Therefore, the drying of the suspension will be high. This indicates that the first measure to reduce evaporation is to reduce the surface area of the sample exposed to the atmosphere. This can be done by reducing the radius of the geometry used and by increasing the depth of the geometry.
- By confining the sample more, the direct exposure of the sample to the atmosphere can be reduced. Consequently, the vapour density in the confined environment increases. This results in reduced evaporation of the suspending medium.
- In some cases, oil can be used to cover the suspension. A small quantity of oil poured over the surface of the suspension can reduce the evaporation. Moreover, the thin layer spread over the suspension does not interfere with the rheological examination.

Suspensions consist of different particles of different sizes, shapes and specific gravity. Unless the particles in the suspension and the suspending medium are of the same specific gravity, unavoidable particle sedimentation or phase separation will occur. Sedimentation or phase separation may result in a discontinuity in the homogeneity of the particles. Moreover, the concentration of particles will be different in different portions. To maintain a constant concentration of particles in the suspension, the fluid in the measuring system is constantly recirculated after a certain time in order to establish a similar SVF. On the other hand, measurement is done in a position where the SVF remains constant.

1.8 RHEOLOGY OF ASPHALT

Asphalt is considered one of the viscoelastic materials. The behaviour of viscoelastic material depends upon the rate and history of loading, temperature, etc. Based on the nature of the material, viscoelastic materials can be classified as viscoelastic solids and viscoelastic liquids. Especially the nature of asphaltic binders significantly depends on the temperature. Subsequently, substantial changes in the mechanical properties of the material can be envisaged based on the change in temperature.

As a general consideration, an increase in temperature decreases the viscosity of asphalt rapidly. Similarly, an increase in the rate of loading increases the complex modulus of the asphaltic binder rapidly. The phase angle, which is a ratio between the loss modulus and the elastic modulus, is 0° for an ideal elastic material; for Newtonian fluids, the phase angle is 90°. Other viscoelastic materials exhibit a phase angle between 0 and 90°. Asphaltic binders of paving grade exhibit a viscoelastic nature, having a phase angle from 45° to 55° at room temperature. The asphaltic binders highly depend on the temperature. Consequently, an increase in temperature increases the viscous nature of the material. Therefore, the viscoelastic nature of the material increases. When the rate of loading is increased, the complex modulus of the material increases at room temperature. This is due to the relative significance of the increased elastic nature of the asphalt. At low temperatures and high rates of loading, asphalt behaves as a viscoelastic solid. On the contrary, for a longer rate of loading and at higher temperatures, asphaltic binders approach Newtonian behaviour (i.e., they behave as a viscoelastic liquid). The phase angle will be close to 90°. The asphaltic materials are viscoelastic in nature. A dynamic shear rheometer (DSR) is used to understand the rheological behaviour of asphaltic materials at different temperatures. Equations 1.18–1.20 are used to calculate the phase angle, storage and elastic modulus. Other than the above parameters, storage viscosity and the Glover–Rowe parameter can be used to relate various performance-related parameters during pavement construction.

Other parameters, such as the stiffness of the asphalt binders, can be measured by a bending beam rheometer. In general, at low temperatures from –30° to 0°, asphalt acts almost like a solid. In those cases, it is important to understand the stiffness. A load is applied to a sample specimen of 100 mm length for a specific time period. The measured deflection is plotted with respect to the applied load. The creep stiffness of the asphalt can be calculated from the load interval of 60 seconds. The creep stiffness of the asphalt should be limited to 300 MPa for the load applied within a time interval of 60 seconds. This limitation is proposed based on the thermal stresses experienced by the pavements at extreme temperatures. Since the asphaltic binders undergo a wide range of processes at different temperatures during pavement construction, it is useful to understand the complex modulus of the asphalt at different temperatures using a frequency sweep. A master curve is constructed by shifting curves at different temperatures with respect to frequency into a single curve. In general, room temperature is selected as a reference temperature, and the other curves are shifted with reference to the room temperature in order to obtain a master curve. The shifting of curves at different temperatures to form a single master curve is done by means of a shifting factor. Applying a shifting factor to a complex modulus vs. frequency curve results in a complex modulus vs. reduced frequency curve. A plot between the shifting factor and temperature gives a relationship between the properties of the asphaltic binder and temperature. In order to characterise asphaltic binders, it is important to have both the modulus and the corresponding phase angle. Phase angle describes the nature of deformation (i.e., whether the material is in a solid or liquid state). The modulus defines the deformation characteristics with respect to applied stress.

Time-temperature superposition theory is considered the underlying theory for the construction of a master curve from DSR data. This theory is based on the fact that the response of a linear viscoelastic material to loading at a high temperature and at a specific frequency is equivalent to the response of a similar material obtained at a low temperature and low frequency. This can be given by different mathematical functions, such as the Williams–Landel–Ferry equation.

1.8.1 Rheological Models: Specific to Asphaltic Binders

Empirical models such as Christensen–Anderson (CA) are used in order to fit the master curve obtained from DSR. The CA model is given by equation 1.28:

$$\left|G^*(\omega)\right| = G_g\left[1+\left(\frac{\omega_c}{\omega}\right)^{\log 2/R}\right]^{-R/\log 2} \tag{1.28}$$

where $G^*(\omega)$ is the complex modulus as a function of reduced frequency, G_g is the maximum complex modulus, ω_c is the frequency that corresponds to the maximum complex modulus, ω is the frequency and R is an empirical parameter that has a good correlation with the relaxation spectra of the material under consideration. A fit of the CA model is shown in Figure 1.17.

Other than the CA model, the generalised Maxwell model is also used to visualise and understand the relaxation spectra of the asphaltic binder. In the Maxwell model, the asphaltic binders are assumed to be dashpots and springs connected in parallel. Each spring and dashpot are associated with a certain magnitude of the spring constant (modulus), viscosity and relaxation spectra. Relaxation time in the Maxwell model can be calculated by dividing the viscosity of the material by the modulus of the material. Relaxation spectra can be calculated by plotting the spring constant with the relaxation time. In general, the plot has a characteristic shape in which the relaxation function increases with respect to relaxation time and decreases, as shown in Figure 1.18.

The characteristic parameters, such as the width and shape of the curve (relaxation function vs. relaxation time), show a direct relationship with the empirical parameter R obtained from the CA model. The wider the relaxation spectrum, the higher the empirical parameter obtained from the CA model, and the curve is more skewed in the shorter relaxation time. The asphaltic materials should have low values of R to have better relaxation properties, i.e., when the stress produced due to a strain decays rapidly, the performance of the asphaltic binder will be good at very low temperatures. Similarly, rheological parameters can be used to understand distress parameters related to asphaltic pavements. Experiments are designed at high temperatures in DSR to understand the rutting and shoving of asphalt. Likewise, fatigue cracking of asphaltic pavements can also be understood by DSR.

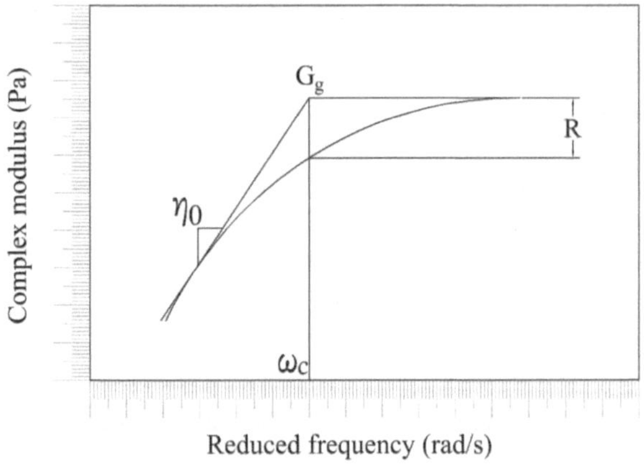

FIGURE 1.17 Christensen–Anderson model.

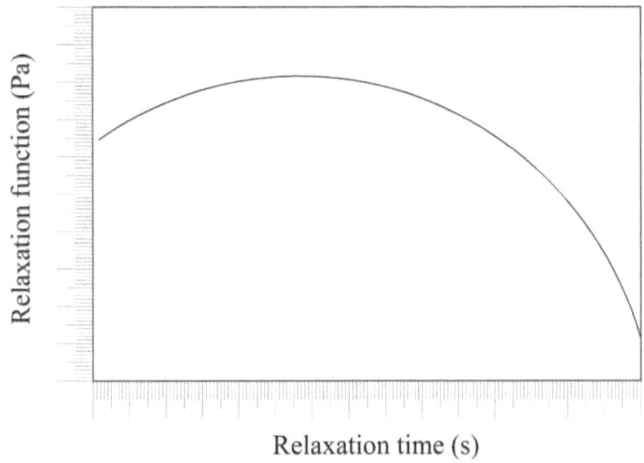

FIGURE 1.18 Relaxation spectrum.

1.9 RHEOLOGY OF COLLOIDAL SUSPENSIONS

Colloidal suspensions exhibit various types of rheological behaviours based on the applied shear, physical characteristics, change in chemical and surface composition, etc. Since most colloidal suspensions are reactive in nature, their microstructure changes constantly with respect to time. Although the definition of a colloid is vague, particles within a size range from nanometres to micrometres are suggested as colloid. The limit on the size was attributed to considering the surrounding medium (suspending medium) as a continuum and to ensuring that the thermal forces are significant. The motion due to thermal forces results in osmotic pressure in a colloidal suspension which is given as shown in equation 1.29:

$$\pi = nk_BT \tag{1.29}$$

where n is the number density of the atoms, k_B is Boltzmann's constant and T is the absolute temperature (K). Movement of a particle of radius 'a' in a suspending medium results in a frictional force or a drag force or a hydrodynamic force acting on the particle. This force is given as shown in equation 1.30:

$$F_H = 6\pi\eta av \tag{1.30}$$

where η is the viscosity of the suspending medium and v is the velocity at which the particle moves in the suspending medium. The Brownian force acting on the particles in terms of thermal energy is given in equation 1.31:

$$F_B = \frac{k_BT}{a} \tag{1.31}$$

The forces will be in the range of femtoNewton (fN). Based on the balance between forces, colloidal particles diffuse if the force due to thermal energy is higher.

A diffusivity constant can be derived for a particle of spherical shape, as shown in equation 1.32:

$$D = \frac{k_B T}{6\pi\eta a} \tag{1.32}$$

A characteristic stress arises due to the force offered by the colloidal particles due to Brownian motion. This characteristic stress is given as shown in equation 1.33.

$$\sigma = \frac{k_B T}{a^3} \tag{1.33}$$

An increase in the particle size decreases the characteristic stress due to Brownian motion. A decrease in the characteristics of stress results in a decreased elastic modulus of the colloidal suspensions. Colloidal suspensions are subjected to gravity; if the force of gravity is higher than the Brownian force, then the particles in the suspension will settle down. However, the magnitudes of the Brownian force and gravitational forces are nearly similar. The effect of gravitational force on the colloidal particles can be calculated by equation 1.34:

$$F = (\rho_P - \rho_m)\frac{4}{3}\pi a^3 \tag{1.34}$$

where ρ_P is the density of the particle, ρ_m is the density of the suspending medium, and a is the radius of the particle.

The particles in the suspension interact via different attributes. For example, the interaction can be due to surface, hydrodynamic forces, dispersion, depletion, etc. The presence of particles induces disturbance in the fluid flow field, which could result in a form of force on the neighbouring particles that are within the range of the flow field. This type of interaction is called hydrodynamic interaction. As a result of fluctuations in the electron cloud of the atoms, dispersion forces arise. Similarly, charge on the surface of the particle results in surface forces. Forces prevailing in the suspension result in either a stable or an unstable suspension. The dispersion forces between the particles arise due to the consequence of the variation in the polarisation of the electron cloud of one atom caused by the fluctuation of the electron cloud of the other atom. In most conditions, variation in the polarisation results in an attractive force. This net effect is known as the London–van der Waals force or dispersion force. The London–van der Waals force is given in equation 1.35:

$$F_{vWF} = -\frac{A(R_1 + R_2)}{(R_1 + R_2)6H^2} \tag{1.35}$$

where A is the Hamaker constant, R_1 and R_2 are the radius of two spheres that are approaching each other and H is the distance between the particles. When the surfaces of two particles approach each other, the van der Waals force becomes negative infinity. This leads to aggregation of the particles in the suspension. Dispersion forces have a longer range; however, they become less effective as the separation distance increases. At zero or infinitesimally small separation distance, the van der

Waals force approaches infinity. Consequently, in order to stabilise the particles in the suspension, the surface charge of the particle imparts forces that avoid agglomeration. Colloidal particles are smaller in size, resulting in a larger surface area. This results in the extremely important nature of surface effects in the case of colloidal dispersions.

Colloidal particles on contact with suspending medium result in surface charges by specific ion adsorption or due to the variation in acidity or basicity of the suspension. The presence of specific surface charges results in the formation of a structured layer. If the surface charges are negative, then the positive charges get accumulated on the surface. The charges in this layer are not movable; this layer is also called the stern layer. The stern layer is followed by a layer of both positive and negative charges, as shown in Figure 1.19, also known as the diffuse layer.

The distribution of charges on the stern and the diffuse layer results in electrostatic potential that decays with respect to an increase in distance from the surface of the particle. The outermost layer beyond the diffuse layer is called as the slip layer. Liquid in the stern, the diffuse and the slip layers are bound to the particle. Beyond the slip layer, the liquid is no longer bound to move along with the particle. The potential at this point is called the zeta potential. The decay in the potential is prominent beyond the stern layer. This decay is given in terms of a decay constant called Debye length k^{-1}. This length depends on the dielectric properties and the ionic strength of the suspending medium. The Debye length is given in equation 1.36:

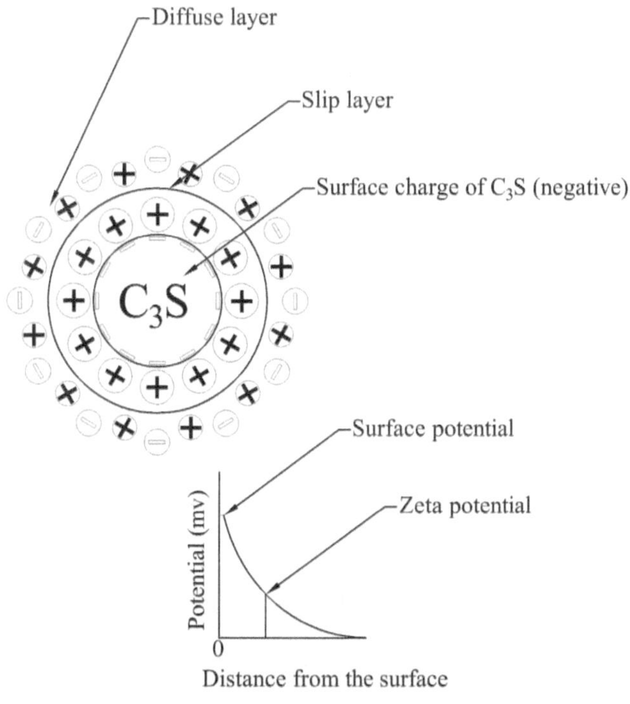

FIGURE 1.19 Double layer formation in C_3S particles (surface charge of C_3S is negative).

$$k^{-1} = \sqrt{\frac{\varepsilon k_B T}{e^2\left(\sum_i z_i^2 n_{i,\infty}\right)}} \tag{1.36}$$

From equation 1.36, we can understand that the Debye length, or the screening length, is inversely proportional to the square of the ion valence (z_i^2). Subsequently, divalent ions will be more effective in screening the electrostatic potential. This effect reduces the electrostatic repulsion, leading to a relatively higher aggregation of the particles when compared to a colloidal suspension with monovalent ions. Similarly, the screening length is inversely proportional to the solution concentration ($n_{i,\infty}$). Sometimes, adsorption of multivalent ions on the surface of the colloidal particles results in charge reversal.

The charge on the particle surface depends on the pH of the suspension. The pH of the suspension plays a crucial role in chemical dissociation and, subsequently, in the electrostatic forces. In some cases, increasing or decreasing the pH may result in charge reversal. The pH at which the charge reversal happens is called the point of zero charge. At the point of zero charge, the suspension possesses only the van der Waals attraction. Due to the lack of stabilisation forces, the particles in the suspension aggregate. Independent research by Derjaguin and Landau and Verwey and Overbeek combined attractive potential due to dispersion with repulsive potential due to electrostatic. The combined potential curve shows a barrier (a primary minimum) and a secondary minimum (due to van der Waals' force) that need to be overcome in order to achieve a stable, well-dispersed suspension. The secondary minimum can be overcome by shearing the suspension. The force applied during the shearing process should be sufficiently high in order to hinder the electrostatic barrier. Although the primary and secondary minimums can be overcome by shear or sonication, their stability is transient and largely depends upon the depth of the primary and secondary minimums, particle surface, the height of the electrostatic barriers and the electrolytic concentration.

1.10 RHEOLOGY OF CEMENTITIOUS SUSPENSION

Cementitious suspensions are considered to be hard-sphere colloidal mixtures. In addition to the forces due to van der Walls attraction, electrostatic repulsion, hydrodynamic force due to shearing, etc., cementitious suspensions are chemically active on contact with water. Hydration of cement results in the formation of products such as calcium silicate hydrate (C–S–H), ettringite, and calcium hydroxide (CH). The formed hydration products develop cohesive forces that eventually lead to development of a hardened network. Although the rheology of cementitious suspensions falls under the category of colloids, their reactivity on contact with water changes their behaviour with respect to time. Moreover, hydration results in temporary as well as permanent structural buildup of cementitious suspensions. In addition, different parameters such as w/b ratio, addition of supplementary cementitious materials (SCMs), and type and dosage of SP affect the rheological behaviour of cementitious suspensions. A short description of the effect of different parameters on the rheological behaviour of cementitious suspension is described in the following sections.

In general, cementitious suspensions are considered to be thixotropic shear-thinning suspensions. On shearing, the apparent viscosity of the suspension decreases; when left undisturbed, a temporary network builds up that results in a thixotropic nature, although based on various parameters, cementitious suspensions could exhibit different behaviours.

1.10.1 Effect of Water-to-Binder Ratio and Solid Volume Fraction

Cement is used as the main ingredient in the manufacture of concrete. Based on the strength and durability requirements, the w/b ratio of the concrete is decided. Though fine and coarse aggregates comprise 60%–70% of the volume of concrete, cement and other SCMs act as a cohesive system to bind the concrete. Consequently, rheological changes in the concrete can be understood by investigating the rheological behaviour of the binder used in the concrete.

In general, an increase in the w/b ratio reduces the SVF; subsequently, the internal resistance offered decreases. The yield stress of the suspension is an inverse function of the w/b ratio. An increase in the w/b ratio reduces the concentration of nearby solid particles, eventually leading to a decrease in the attractive potential between the particles due to the effect of dilution. On the other hand, an increase in the w/b ratio increases the hydration process since the amount of water available per fixed volume is high. Therefore, the formation of temporary or permanent networks with respect to time increases. This leads to an increased thixotropic nature of the suspensions. A decrease in the w/b ratio (increase in SVF) results in increased internal resistance between the particles. This leads to increased apparent viscosity and yield stress in the suspension. As the w/b ratio decreases, the contact between the particles enhances, which leads to a very high internal resistance of the suspension. Moreover, the yield stress of the suspension also increases since the probability of finding a nearby particle will be higher, which results in easy formation of networks.

1.10.2 Effect of Superplasticiser (SP) on Rheology

Cementitious suspensions are attractive in nature due to the presence of opposite charges that evolve from multi-mineral compounds in the system during hydration. Tri-calcium silicate in the system acquires a negative charge, and tri-calcium aluminate in the system acquires a positive charge due to the ionisation at high pH of the system. The transient availability of opposite charges and the high ionic concentration of the suspension result in the suspension's overall attractive potential. This leads to agglomeration of the particles; agglomerated particles entrap a significant volume of water. This leads to reduced water availability for the dispersion, which consequently leads to a higher yield stress in the suspension. Therefore, in order to disperse the particles more uniformly, dispersants in the form of plasticisers and SPs are added. Based on the dispersive forces provided by the dispersants, dispersants are classified as high-range, mid-range and low-range water reducers. SPs are designed to be high-range water reducers since they act based on both electrostatic repulsion and steric hindrance. SPs ionise on contact with cementitious suspension. SPs are poly-acrylates and poly-methacrylates, and ionisation of acrylates or methacrylates results in negative charges. The negatively charged SPs get attracted to the surface of cement particles and provide repulsive potential through steric hindrance. This leads to release of entrapped water. The release of entrapped water reduces the yield stress and the internal resistance. Investigation on the usage of SP on dispersion show that increased usage of SP can increase the shear-thickening behaviour of the suspension.

1.10.3 Effect of Supplementary Cementitious Materials on Rheology

SCMs are added to enhance the properties and produce sustainable concrete. Based on the physical properties and the chemical composition of the added SCM, the rheological behaviour of cementitious suspension differs. Some of the most widely used SCM are ground granulated blast furnace slag (GGBS), FA, micro silica, agro-based residues such as rice straw ash, rice husk ash, sugarcane bagasse-ash, etc.

Most of the SCMs added as a replacement have a relatively low specific gravity compared to OPC. Subsequently, substitution of OPC by SCMs on a weight basis results in greater volume. Enhanced volume results in the requirement of increased water to have similar rheological behaviour as that of suspension only with OPC. Although the substitution of OPC by SCMs on a volume basis results in a constant SVF, other factors such as physical, chemical, and surface characteristics alter the rheological behaviour significantly.

Partial substitution of OPC by GGBS increases the apparent viscosity of the suspension considerably. GGBS, obtained as a byproduct during the extraction of iron, is a latent hydraulic material. The slag obtained during the process is highly irregular in shape and has a relatively higher surface area. Highly irregular shape and surface area result in excess volume of water to cover the particles. Subsequently, the physical and surface characteristics of GGBS increase the apparent viscosity and the yield stress. GGBS is a latent hydraulic material; consequently, substitution of OPC by GGBS reduces the reactivity. A reduction in reactivity reduces the formation rate of temporary/permanent networks. Therefore, the thixotropic nature of the suspension decreases. Similarly, due to the reduced reactivity, de-agglomeration of the particles in the suspension is easier. This leads to an early-onset and higher shear-thickening intensity.

Based on its chemical composition, FA is classified either as class F or as class C. Class F FA are considered to be pozzolanic in nature (i.e., the ability to react with CH to form C–S–H), and class C FA are considered to be latent hydraulic. Substitution of OPC by FA reduces the yield stress and the apparent viscosity of the suspension because of their spherical shape. However, the usage of SP in order to improve the rheological behaviour of the suspension may highly depend upon the source of FA used since, the presence of unburnt carbon in the FA might affect the adsorption ability and dispersion of FA-based suspensions.

The use of micro silica as a substitution for OPC is generally restricted to 15% by weight. This restriction is based on the fact that the surface area of micro silica is enormously high. Enormous surface area results in high apparent viscosity and yield stress.

Several agro-based ashes are used as a substitute for OPC. Usage of agro-based residues abruptly increases the SVF. A significant increase in the SVF results in increased internal resistance. Moreover, the yield stress of the suspension increases relatively when compared to OPC-based suspensions. SP can be used to reduce the yield stress and internal resistance of the agro-based suspensions; however, the presence of unburnt carbon in the ashes results in incompatibility. Primarily, the presence of unburnt carbon results in unexpected absorption of SP. This results not only in poor performance of SP but also in an abrupt increase in yield stress and apparent viscosity of the suspension.

1.11 RHEOLOGY OF CONCRETE

Concrete is a mixture of polydisperse particles (aggregates) suspended in a non-Newtonian suspension (cementitious suspension). The presence of aggregates results in a major challenge in predicting the rheological properties of concrete. The viscosity of the diluted Newtonian suspensions depends on the viscosity of the suspending medium and the volume fraction of the particles in the suspension. This is given by equation 1.37:

$$\eta_{\text{diluted}} = \eta\left(1 + 2.5\phi\right) \tag{1.37}$$

where η_{diluted} is the viscosity of the suspension, η is the viscosity of the suspending medium and ϕ is the SVF.

Interaction between the particles in the suspension will be minimal or negligible for a diluted suspension. As the SVF increases, hydrodynamic, inertial and frictional interactions increase. When the SVF reaches a maximum value ϕ_m, the viscosity of the suspension reaches infinity. Once the SVF reaches a maximum value ϕ_m due to the formation of a percolation network, shearing of the suspension becomes difficult. Based on the deformation range and ratio between the volume of the suspended phase and that of the volume of suspension, a general classification of interactions and their dominance over a varied shear deformation range and SVF for pastes and granular suspensions can be obtained. This classification can act as a general framework to classify the influential parameters that may determine the behavioural changes of concrete.

Based on the equilibrium between interactions and dimensional analysis, certain dimensionless numbers can be arrived. They are Peclet number, Flow number and Reynolds number, which are, respectively, the ratios between forces due to hydrodynamic interaction and Brownian interaction, hydrodynamic interaction and colloidal interaction, and inertial interaction and hydrodynamic interaction. In some interactions, especially in granular (conventional vibrated concrete [CVC]) and granular pastes (self-compacting concrete [SCC]), neglecting hydrodynamic and soft interactions, certain specific regime transitions can be observed that correspond to frictional and collisional regimes. These regimes can be characterised based on the ratio between Strokes number and Leighton number.

These numbers are the ratios between the characteristic kinetic energy of the particle and that of characteristic viscous energy and the characteristic frictional energy to that of characteristic viscous energy, respectively. The corresponding ratio between Strokes number and Leighton number is called the Bagnold number, and the magnitude of the Bagnold number characterises the transformation of the regime from frictional to collisional. Their ratios correspond to the equilibrium state between two or more interactions. This ratio can be very useful in determining deformation ranges, where the behavioural changes in the respective granular suspension can be ascertained based on their relationship to fundamental interactions.

CVC and SCC have a high fraction of fine and coarse aggregates suspended in paste. Their interaction will be typically frictional or collisional. In general, the yield stress of the CVC depends upon the w/b ratio, type and content of paste, proportion, distribution of size and shape of the aggregates being used, etc. An increase in the w/b ratio decreases the yield stress. Similarly, a decrease in the proportion of coarse aggregate decreases the inter-particle friction. A decrease in the inter-particle friction decreases the yield stress. An increase in the paste content increases the viscosity. In the case of SCC, the fraction of coarse aggregate used is less. Moreover, SCCs are designed to have low or very low yield stress. Although SCCs are designed to have low yield stress, because of their high powder content, their relative viscosity is higher. This is highly attributed to the very low inter-particle distance, which leads to relatively higher hydrodynamic dissipation.

Binders in the concrete hydrate when they come into contact with water. Hydration results in the thixotropic nature of the concrete. Based on the type and percentage replacement of SCMs used, the thixotropic nature of the concrete varies. For example, the use of micro silica results in a higher thixotropic nature of the concrete because of its early reactivity. SCC in general has a relatively higher thixotropic nature when compared with CVC. SCC of moderate to high thixotropic nature is preferable during the casting process. High thixotropic nature results in faster buildup of the percolation network, which leads to less pressure exerted on the formwork. On the other hand, SCC with poor thixotropic nature results in high pressure exerted on the formwork during the casting process, leading to failure of the formwork and additional costs during the process of construction. From the process point of view, especially during pumping or the process that involves high shear, shear-induced particle migration takes place inside the pipe, which leads to formation of different layers across the cross section of the pipe. The lubrication layer that is formed at the interface of the pipe and the concrete facilitates the pumping process. Disruption of the layer results in inefficient pumping (Moorthi et al., 2023). An increase in the yield stress of the layers results in the high pressure required for the piston to pump the concrete, which is especially the case with CVC. CVC has a predominantly high yield stress because of the large coarse aggregate proportion, which leads to enhanced inter-particle friction between the particles in the suspension. In the case of SCC, the yield stress is minimal; however, high powder content results in increased internal resistance, thereby exhibiting shear-thickening behaviour during pumping. Besides, the high thixotropic nature of SCC needs to be kept under control in order to have a smooth and flawless pumping process.

BIBLIOGRAPHY

ASTM D4440. (2015). "Standard Test Method for Plastics: Dynamic Mechanical Properties Melt Rheology." *American society for testing and* materials: 1–5.

ASTM D4473. (2021). "Standard Test Method for Plastics: Dynamic Mechanical Properties: Cure Behavior." *American society for testing and* materials: 1–6.

Coussot, P. (2005). *Rheometry of Pastes, Suspensions, and Granular Materials Applications in Industry and Environment.* 1st ed. John Wiley & Sons, Inc, Hoboken, NJ.

Mezger, Thomas G. (2006). Instruments. In *The Rheology Handbook* (pp. 234–242). Vincentz Network, Hannover, Germany.

Moorthi, P.V.P., Pra Mio, F., Nanthagopalan, P., Ferrara, L. (2020). "Onset and Intensity of Shear Thickening in Cementitious Suspensions – A Parametrical Study." *Construction and Building Materials*: 244, 118292, 1–15.

Moorthi, P.V.P., Athira Gopinath, Nanthagopalan, P. (2023). "Mechanistic Origins of Concrete Pumping: A Comprehensive Outlook and Way Forward." *Magazine of Concrete Research*: 75 (7), 353–366.

MULTIPLE CHOICE QUESTIONS

1. Ideally, those materials that can regain their original shape on release of the applied stress is called as ....
   a. Viscoelastic liquids
   b. Solids
   c. Viscoelastic solids
   d. None of the above
2. Attractive forces in the suspension are attributed due to ....
   a. Van der Waals' forces
   b. Hydrogen bonding
   c. Ion-correlation forces
   d. All the above
3. Presence of high zeta potential results in ....
   a. Most stabilised suspension
   b. Highly unstable suspension
   c. Moderately stable suspension
   d. None of the above
4. Low zeta potential results in ....
   a. Agglomeration of particles
   b. De-agglomeration of particles
   c. Stable suspension
   d. All the above
5. Formaion of percolation network due to flocculation of the particles in the suspension results in ....
   a. Shear-thinning behaviour
   b. Shear-thickening behaviour
   c. Viscoelastic behaviour
   d. All the above

6. The innate ability of the particles in the suspension to resist stress is called as _________
   a. Static yield stress
   b. Dynamic yield stress
   c. Plastic viscosity
   d. Apparent viscosity
7. For a Newtonian suspension, the absolute resistance offered by the suspension for an external stress is _________
   a. Linearly dependent
   b. Non-uniformly
   c. Constant
   d. None of the above
8. Which of the following is not a non-Newtonian material _________
   a. Honey
   b. Cement paste
   c. Bentonite suspension
   d. All the above
9. For a suspension at rest, if the attractive forces is more than the repulsive forces. How the behaviour of the suspension can be described when sheared?
   a. Suspension will possess yield stress
   b. Suspension will possess yield stress and shear thinning in nature
   c. Suspension will be flocculated in nature and shear thickens
   d. All the above
10. Re-aggregation of particles in the suspension due to shear is called as _________
    a. Agglomeration or flocculation
    b. Shear thinning
    c. Thixotropy
    d. Hydroclusters
11. The suspension whose behaviour is time dependent and shear thinning in nature is called as _________
    a. Anti-thixotropic
    b. Rheopectic
    c. Thixotropic
    d. None of the above
12. **Statement A:** If the motor and the transducer are combined and mounted in a side of the rheometer.
    Statement B: Then only shear stress can be given as input. Shear rate cannot be given as input.
    a. Statement A is in line with Statement B
    b. Statements A and B are contradictory in nature
    c. Both the statements are incorrect
    d. None of the above

13. An encoder is used as _________
    a. Transducer
    b. Electronic controller
    c. Position sensor
    d. None of the above
14. Why tachogenerator cannot be used in case of static test?
    a. The measurement needs to be conducted at low deflection angle
    b. They work based on inverted electrometer
    c. Both a and b
    d. None of the above
15. For a low-viscous material _________ is the preferred measuring system.
    a. Parallel plate
    b. Cone and plate
    c. Vane
    d. Concentric cylinders
16. Why gap between the cup-and-bob measuring system should be maintained minimum?
    a. In order to minimise the material being used
    b. So that uniform stress is being applied throughout the sample
    c. To avoid instabilities such as secondary flow and turbulent flow
    d. All the above
17. Which of the following methods can be used to reduce 'Taylor vortices' in co-axial measuring system?
    a. Searle method
    b. Couette method
    c. Both a and b
    d. None of the above
18. In a co-axial measuring system, sensitivity of the measuring system at low torque value _________ When the radius of the inner cylinder or the bob increases.
    a. Decreases
    b. Increases
    c. Remains the same
    d. None of the above
19. Which of the following measuring system can be used for sample of high flowability?
    a. Parallel plate
    b. Cone and plate
    c. Co-axial measuring system
    d. All the above
20. For which of the following system, shear rate is constant across the radius of the system?
    a. Parallel plate
    b. Cone and plate
    c. Co-axial measuring system
    d. None of the above

21. Why shear rate is constant across the radius of the system for a cone-and-plate measuring system?
    a. Conversion constant is inversely proportional to the radius of measuring system
    b. Conversion constant is proportional to the cone angle of measuring system
    c. Conversion constant is independent of the measuring system geometry
    d. None of the above
22. Why low cone angle is preferred in cone-and-plate measuring system?
    a. In order to achieve high angular velocity
    b. In order to achieve high shear rate
    c. Both a and b
    d. None of the above
23. In a parallel plate system, the sensitivity of the shear stress at low values _________when the radius of the plate increases.
    a. Decreases
    b. Increases
    c. Remains constant
    d. None of the above
24. In a parallel plate system, shear rate is minimum at _________
    a. the edge of the plate
    b. the centre of the plate
    c. at $1/3^{rd}$ of the radius of the plate
    d. at $2/3^{rd}$ of the radius of the plate
25. Shear rate in a parallel plate system can be increased by _________
    a. Increasing the radius of the plate
    b. Decreasing the gap between the plates
    c. Both a and b
    d. None of the above
26. In a parallel plate system, for a multi-disperse suspension, the particles in the suspension migrate towards _________ When shear rate is increased.
    a. Towards the edge of the plate
    b. Towards the centre of the plate
    c. Remain as such in their position
    d. None of the above
27. What happens to the viscosity of the suspension if the viscosity of the suspending medium increases _________
    a. Increases
    b. Decreases
    c. Remains same
    d. Increases and decreases
28. Increase in solid volume fraction results in _________
    a. Early onset of shear thickening
    b. Increase in viscosity of the suspension
    c. Both a and b
    d. None of the above

29. Suspension A has particles that are of spherical in nature with a solid volume fraction of 0.450. Suspension B has particles that are of non-spherical in nature with similar solid volume fraction as that of suspension A. Which of the following statements hold true?
    1 Suspension A has relatively higher viscosity than Suspension B
    2 Specific surface area of the particles in Suspension B is higher than the specific surface area of the particles in Suspension A
    3 Viscosity of Suspension B is higher than Suspension A
    4 Specific surface area of the particles in Suspension A is higher than the specific surface area of the particles in Suspension B
    a. 1 and 3
    b. 1 and 2
    c. 3 and 4
    d. 2 and 3
30. Yield stress of the Suspension A is higher than the yield stress of Suspension B. Which of the following statement is/statements are correct?
    a. Suspension A shear thickens earlier
    b. Suspension B shear thickens earlier
    c. Both the suspensions shear thickens at same shear rate or shear stress
    d. None of the above
31. At similar w/b ratio, if OPC in the system is replaced by fly ash (FA). Which of the following remains true? Assuming that the replacement is 25% by volume.
    a. Yield stress of suspension with FA will be higher than the suspension with only OPC
    b. Yield stress of suspension with FA will be lower than the suspension with only OPC
    c. There won't be any predominant change in the yield stress of the suspension
    d. None of the above
32. If OPC is replaced with less reactive, low specific gravity material by weight. What will be the thixotropic nature of the suspension?
    a. Increases
    b. Decreases
    c. Remains the same
    d. None of the above
33. Which of the following behaviour will be predominant on shearing due to the addition of solids that have higher aspect ratio?
    a. Shear thinning
    b. Shear thickening
    c. Both a and b
    d. None of the above
34. Why sample needs to be sheared prior to rheological testing?
    a. To circumvent significantly non-reproducible rheological data
    b. For the internal structure of the sample to reach a steady state
    c. To avoid uncontrolled state of the sample
    d. All of the above

35. For a viscoelastic material, for a stress less than the yield stress, the shear rate experience by the material is _________
    a. More than zero
    b. Equal to zero
    c. Less than zero
    d. None of the above
36. Which of the following holds true for a Bingham material?
    1 Bingham material is a type of non-Newtonian material with a yield stress
    2 Bingham material can exhibit shear-thinning and shear-thickening behaviours
    3 Bingham materials are Newtonian in nature
    a. Both 1 and 2
    b. Only a
    c. Both 2 and 3
    d. 1, 2 and 3
37. Non-Newtonian material behaviour can be represented by a power law. $\tau = a\dot{\gamma}^n$. Here, τ is the shear stress, a is the flow index and n is the power index. If the material is shear thinning in nature, which of the following value is/values are correct?
    a. 1.0
    b. 0.98
    c. 1.10
    d. 1.01
38. Consider the following equations: $\tau = G\gamma_0 \sin\omega t + \mu\gamma_0\omega\cos\omega t$ and $\tau = \tau_0 \sin(\omega t + \delta)$. Which of the following is/are correct?
    a. $\tau_0 \cos(\delta) = G\gamma_0$
    b. $\tau_0 \sin(\delta) = \mu\gamma_0\omega$
    c. $\tan(\delta) = \dfrac{\mu\omega}{G}$
    d. All the above
39. Solid nature and viscous nature of the material can be monitored by SAOS test. Which of the following is/ are true?
    1 Loss factor is the ratio between storage modulus and loss modulus
    2 Loss factor is the ratio between loss modulus and storage modulus
    3 Storage modulus is used to understand the liquid behaviour and loss modulus is used to understand the solid behaviour
    4 Storage modulus is used to understand the solid behaviour and loss modulus is used to understand the liquid behaviour
    a. 1 and 3
    b. 2 and 3
    c. 1 and 4
    d. 2 and 4

40. Shear-induced particle migration results in _________
    a. Phase separation
    b. Formation of slip layer
    c. Turbulence
    d. Inertial effect

1 b	2 d	3 a	4 a	5 c	6 a	7 c	8 a	9 b	10d
11 c	12 b	13c	14 c	15 d	16d	17b	18b	19c	20b
21 c	22 b	23b	24 b	25 c	26a	27a	28c	29d	30b
31 b	32 b	33a	34 d	35 b	36b	37b	38d	39d	40b

2 X-ray Diffraction

2.1 OVERVIEW

X-rays were discovered by the German physicist Roentgen in 1895. The nature of X-rays was unknown at the time of discovery, and hence the name X-rays. The first use of X-rays was in the field of medical sciences. The relative absorption of X-rays by matter as a function of its density and the average atomic number was the basis for using X-rays to develop diagnostic methods. The same property of X-rays also finds application in other industries as a non-destructive testing method to identify defects such as internal cracks and flaws in materials. Later, in 1912, Max von Laue confirmed the wave nature of X-rays by carrying out experiments on X-ray diffraction (XRD) using a copper sulphate crystal. He stated that each atom acts as an X-ray scattering point, and diffraction of X-rays is possible if a crystal has a long-range periodic repetition of lattice planes, provided the X-ray wavelength is in the range of the interatomic distance of the crystal. His experiment paved the way for the development of X-ray crystallography, a widely accepted technique to study the properties of crystalline materials. The X-ray crystallographic study involves collecting and processing the X-rays diffracted by the material to generate an XRD pattern. The XRD pattern consists of characteristic peaks of different intensities at specific diffraction angles. The position of diffraction peaks is unique for a material since it depends on the interplanar spacing of the crystal lattice in the material. The XRD method gives rapid and accurate data on the crystal structure's characteristics. Bragg is known as the first person to have determined crystal structures using the XRD technique. He defined the conditions for XRD by a crystal in a simpler mathematical form than von Laue and successfully determined the structures of NaCl, KCl, KBr and KI crystals.

This chapter gives a brief explanation of the terminologies related to the XRD technique and the components of an X-ray diffractometer. The various types of interaction of X-rays with the specimen, such as absorption, scattering and diffraction, are discussed. The factors influencing XRD data, including specimen characteristics, instrument parameters and operating conditions, are also explained in detail. An in-depth explanation of the working principle of XRD and a stepwise procedure for conducting the experiment and data processing are given in the subsequent sections. The application of XRD in various civil engineering fields is briefly described in this chapter. A section focusing on the issues related to characterisation using the XRD technique is also presented.

DOI: 10.1201/9781032635392-2

2.2 TERMINOLOGIES

Amorphous materials	Materials with crystallites consisting of non-identical unit cells with short-range periodicity are called amorphous materials.
Crystal lattice	The crystal lattice is a periodic arrangement of atoms or molecules in a three-dimensional space. The atoms may be distributed at equal or different repeat intervals in three orthogonal directions. In a cubic system, the atoms are distributed at equal intervals along the three directions.
Crystalline materials	Materials with crystallites comprising a highly ordered long-range arrangement of unit cells are known as crystalline materials.
Crystallite	Several unit cells are systematically arranged or grouped together to form a crystallite.
Crystal structure	A series of parallel-sided unit cells are closely packed in three orthogonal directions to form a crystal structure.
Diffraction	Diffraction is the phenomenon of bending of electromagnetic (EM) radiation as it passes around the edge of an obstacle or aperture.
d-spacing	The distance between two parallel lattice planes is known as interplanar spacing, or *d*-spacing. It is denoted as d_{hkl} or d. The interplanar spacing is different for different sets of lattice planes in a crystal.
Interference	Interference is defined as the phenomenon in which two or more waves superpose to produce a resultant wave whose amplitude is the algebraic sum of the amplitudes of individual waves.
Lattice parameters	The edges of a unit cell along the three orthogonal axes (a, b, c) and the three interaxial angles (α, β, γ) are termed the lattice parameters. They are used to describe a unit cell.
Lattice plane	Crystallographic planes, or lattice planes, are equally spaced parallel planes containing atoms or groups of atoms.
Miller indices	The Miller indices represent a set of parallel lattice planes and are proportional to the reciprocals of the fractional intercepts that the plane makes with the crystallographic axes. The Miller indices are denoted as an integer triplet *hkl*, where $1/h$, $1/k$ and $1/l$ are the fractional intercepts.
Phase shift	When two identical waves do not coincide, they are said to have a phase shift or phase difference. The phase shift can be measured on a linear scale (Δ) or an angular scale ($\delta\varphi$).
Scattering	Scattering is the phenomenon of re-emission of incident energy absorbed by a material.
Step size	The difference between two consecutive diffraction angles considered for data collection in an XRD study is known as the step size.
Unit cell	A unit cell is a periodically repeating fundamental unit of a crystal formed by the bonding of atoms in a space lattice.
Weiss indices	The Weiss indices for a lattice plane are calculated as the intercept made by the plane on the crystallographic axis (ma, nb, and pc) divided by the unit length on the same axis (a, b, and c). However, when a lattice plane is parallel to the crystallographic axis, the Weiss indices become infinity. Therefore, it is more practical to represent a plane using Miller indices.
X-ray brightness	The number of X-ray photons emitted per unit area per second when a high-energy electron bombards a target metal is defined as the X-ray brightness.

X-ray intensity	The intensity of an X-ray is defined as the rate of energy flow through a unit area perpendicular to the direction of propagation. It is higher when X-rays are focused on a smaller area of the material.
X-ray spectrum	An X-ray spectrum consists of characteristic spectral lines or peaks of varying intensity. Each peak corresponds to X-rays diffracted from a specific set of lattice planes in the crystal. The characteristic peaks in the spectrum are superimposed by a broad range of continuous X-rays known as bremsstrahlung, or white radiation.

2.3 NATURE OF X-RAYS

X-rays are high-energy EM radiation. All the EM radiations exhibit dual wave-particle characteristics, i.e., they have a wave nature and a particle nature. EM radiations consist of oscillating electric fields and magnetic fields that are orthogonal to each other. In the EM spectrum, X-rays fall in the region between gamma rays and ultraviolet rays. According to classical theory, X-rays are characterised by wave motion, and their wavelength is measured in angstrom (1 Å = 10^{-10} m) or *X* units. They have a short wavelength in the range of 0.1–100 Å. Considering the particle nature of X-rays, they are made up of packets of energy called X-ray photons. In terms of photon energy, X-rays lie in the range of 0.1–100 keV. The relation between energy E and wavelength λ of X-rays is given as shown in equation 2.1.

$$E = h\upsilon = hc/\lambda \tag{2.1}$$

where h is Planck's constant, υ is the frequency of X-rays, and c is the speed of light. The equation shows that the energy of X-rays is inversely proportional to their wavelength.

2.4 TYPES OF X-RAYS

Generally, two types of X-rays contribute to an X-ray spectrum: X-ray continuum and characteristic X-rays (shown in Figure 2.1).

2.4.1 X-RAY CONTINUUM

In an X-ray tube, the electrons produced from a cathode strike the target metal at a very high velocity. However, a few electrons may deviate from their path and decelerate in the presence of a strong electric field near the atomic nucleus of the target metal. During this process, most of its kinetic energy is converted into heat, and less than 1% is released as X-ray energy. The X-rays produced due to the deceleration of the incident electrons are known as bremsstrahlung X-rays. As these X-rays have a continuous spectrum, they are also called X-ray continuums or continuous X-rays. The X-ray continuum is made up of radiation of different wavelengths similar to white light, and hence, it is also known as white radiation. The nature of the atoms on which the electrons strike has no role in determining the characteristics of the X-ray continuum.

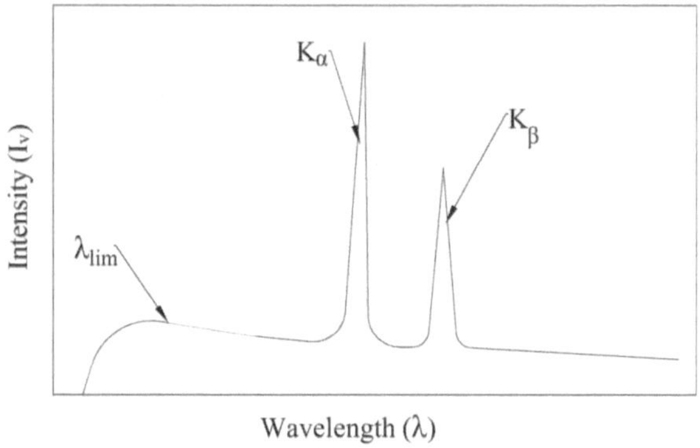

FIGURE 2.1 X-ray spectrum showing characteristic and continuous X-rays.

If an electron loses its energy instantaneously on striking an atom, the X-rays produced will have the maximum photon energy or the shortest wavelength. The wavelength corresponding to maximum energy is known as the short wavelength limit (λ_{lim}). If the wavelength of X-rays is shorter than the short wavelength limit, the intensity of the X-ray continuum will be zero. The short wavelength limit depends on the potential difference applied in the X-ray tube.

2.4.2 Characteristic X-rays

When the energy of electrons incident on the target metal is sufficiently high, it may overcome the binding energy of a tightly bound electron in the atom. This causes the ejection of the electron from the target metal. An atom has different energy levels with unique binding energy, E_K, E_L, E_M, etc. The binding energy of electrons decreases as the distance of the electron from the nucleus increases ($E_K > E_L > E_M$). The free electron leaves the atom with a kinetic energy equal to the difference in energy between the incident electron E and the binding energy of the atomic electron. The release of the inner shell electron creates a vacancy in the corresponding shell (say, the K shell) and leaves the atom in an excited or ionised state. The vacancy is filled by the transition of an electron from an outer shell (L or M shell) to the vacant inner shell (K shell). The electron transition is accompanied by the simultaneous release of energy in the form of X-rays. This brings the atom from the ionised state to the ground state. The radiation leaves the atom with an energy equal to the difference in energy between the two shells (E_K–E_L, E_K–E_M). Since the radiation energy is element-specific, these radiations are known as characteristic X-rays. They are seen as sharp and narrow peaks in the X-ray spectrum and are also called characteristic spectral lines.

The characteristic X-rays are categorised into various spectral line series, such as the K series, the L series, etc., based on the atomic shell in which the electron transition occurs. If an L shell electron fills the vacancy in the K shell, it is called $K\alpha$ radiation, and if the electron transition is from the M shell to the K shell, it is known as $K\beta$ radiation. Figure 2.2 shows the emission of characteristic X-rays from the

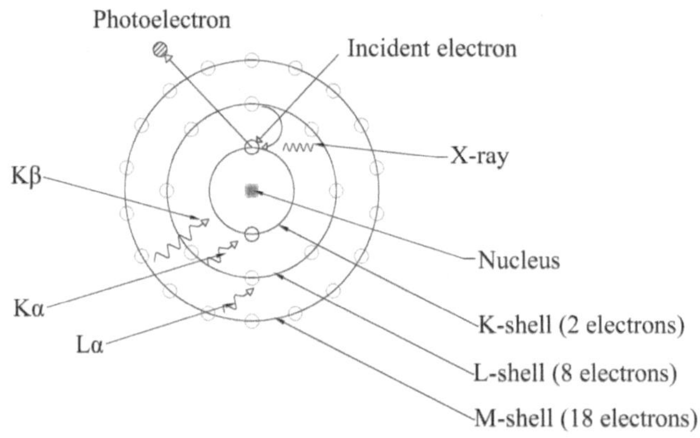

FIGURE 2.2 Characteristic X-rays emitted from the target metal atom.

target metal atom when incident electrons strike the atom. The probability of filling a vacant *K* shell with an electron from an *L* shell is higher than that from an *M* shell. Therefore, the $K\alpha$ spectral lines are always stronger than the $K\beta$ lines. Similar to the *K* series, electron transitions from the outer shells to the vacant *L* shells lead to the generation of spectral lines in the *L* series.

K, *L*, and *M* correspond to the shell with the principal quantum numbers, $n = 1$, 2, and 3, that specify the main energy level or orbit. Since the longer wavelengths are easily absorbed, mostly $K\alpha$ and $K\beta$ radiations are used for XRD studies. The $K\alpha$ radiation consists of two components in the X-ray spectrum, namely, $K\alpha_1$ and $K\alpha_2$. Even though the spectral lines of $K\alpha_1$ and $K\alpha_2$ lie close to each other, they can be separated using several techniques, and when resolved into separate peaks, they are known as $K\alpha$ doublet. $K\alpha_1$ and $K\alpha_2$ radiations correspond to electron transitions from $2p_{1/2}$ to $1s_{1/2}$ and $2p_{3/2}$ to $1s_{1/2}$, respectively, where s ($\ell = 0$) and p ($\ell = 1$) represent the angular quantum number ℓ. The angular quantum number gives information about sub-energy levels, or orbitals. The subscripts 1/2 and 3/2 are the total angular momentum quantum numbers j. The total angular momentum quantum number is the sum of the angular quantum number and the spin quantum number s. For the s orbital, $\ell = 0$, which implies $j = 1/2$ as $s = \pm 1/2$. Similarly, for the p orbital, $\ell = 1$ and $j = 1/2$ or 3/2. The $K\beta$ radiations have comparatively shorter wavelengths and lower intensities than the $K\alpha$ radiations. Similar to the $K\alpha$ radiations, the $K\beta$ radiations have several characteristic lines, with $K\beta_1$ and $K\beta_3$ being the strongest. However, these spectral lines lie so close to each other that they are not easily distinguishable in the spectrum. In addition to $K\alpha$ and $K\beta$, there are other spectral lines corresponding to electron transitions between different energy shells that are not considered in the XRD analysis due to their low intensity. Among all the spectral lines, $K\alpha_1$ has the highest peak intensity, followed by $K\alpha_2$ (half of $K\alpha_1$) and $K\beta$ (one-fifth of $K\alpha_1$).

2.5 INTERACTION OF X-RAYS WITH A SPECIMEN

The penetration depth of an X-ray beam into a specimen depends on the incident energy, the incident angle, the absorption coefficient, and the elemental composition of the material. An X-ray beam propagating through a specimen is partly absorbed and partly transmitted. The absorption of X-rays by a material is attributed to two phenomena: true absorption and scattering. The true absorption of incident X-ray energy is caused by electron transitions within the atom. When the absorbed energy is greater than the excitation potential of the atomic electron, an electron is ejected from the atomic shell. This process is known as the ionisation of an atom, and the ejected electron is called a photoelectron. The characteristic X-ray emitted in the process is called fluorescent radiation. In addition to the scattering of the incident X-rays, they may be transmitted through the material without any energy loss; but the trajectory might change as they pass through the material.

2.5.1 Absorption of X-rays

The absorption of incident X-rays by material causes attenuation of the X-ray intensity. The attenuation of incident energy is proportional to the distance travelled through the material. For an X-ray of intensity I_o travelling through a distance dx, the relation between the intensity of the transmitted X-ray and the distance travelled is given in equation 2.2:

$$\frac{I}{I_o} = e^{-\mu x} \tag{2.2}$$

where μ is the proportionality constant called the linear absorption coefficient of the material and $e^{-\mu x}$ is called the attenuation factor. The same equation is rewritten in a more convenient form, as shown in equation 2.3:

$$I = I_o e^{-\left(\frac{\mu}{\rho}\right)\rho x} \tag{2.3}$$

The quantity μ/ρ is called the mass absorption coefficient. The mass absorption coefficient is independent of the chemical and physical state of a material. Hence, it is characteristic of an element at a specific wavelength. The variation of absorption coefficient as a function of wavelength gives an idea about the absorption characteristics of a material. A plot of μ versus λ is shown in Figure 2.3. As depicted in the plot, as λ decreases, μ steadily decreases until a value of λ is reached, after which an abrupt increase in μ` is observed. Then, μ continues to decrease on further reduction of λ. Thus, the plot of the absorption coefficient of an element as a function of the wavelength of X-rays has two continuously decreasing portions separated by a sharp discontinuity at a distinct wavelength. The continuous branches in the μ versus λ plot represent the absorption of energy due to X-ray scattering by electrons in the atoms. The point at which the sudden discontinuity of μ occurs is known as the critical absorption wavelength or absorption edge. The absorption edge represents the limiting point at which X-rays have sufficient energy to eject an electron from the atom. The same phenomenon occurs for other electrons in the atom at various wavelengths, resulting in multiple absorption edges. In the continuously decreasing portions, μ is proportional to the third power of the product of Z and λ, where Z is the atomic number of the element, and at the point of discontinuity, μ changes by a factor of 6–8.

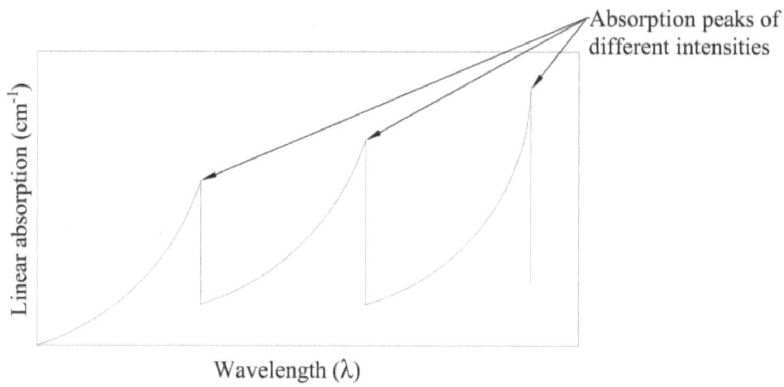

FIGURE 2.3 Linear absorption coefficient as a function of wavelength.

The energy of X-rays is inversely proportional to their wavelength, which means that longer-wavelength X-rays have lower energy. The interaction of low-energy X-rays with material does not lead to the ionisation of the atom. As the wavelength decreases, the energy increases, and at the critical absorption wavelength, the energy will be just sufficient for the ejection of an electron from its energy shell. At this point, maximum absorption of X-ray energy occurs, thereby overcoming the electron binding energy. This point is known as the absorption edge. On the shorter-wavelength side of the absorption edge, the X-ray energy is well above the electron's binding energy. As the wavelength decreases, the chance of X-ray transmission without getting absorbed is higher. Conversely, on the longer-wavelength side of the absorption edge, X-ray absorption is lower due to insufficient energy of the X-rays to eject an electron from the atom.

2.5.2 Scattering of X-rays

Scattering is the phenomenon of re-emission of incident energy absorbed by a material. A collimated beam of X-rays striking a crystal interacts with the atoms in the crystal. It causes the electrons in the atom to oscillate about their mean position at the same frequency as the incident X-rays. The acceleration and deceleration of electrons occur due to the influence of the alternating electric field component of the X-rays. According to classical EM theory, an accelerating charged particle such as an electron emits EM radiation with the same frequency. In the case of crystals, the emitted EM radiation is in the form of X-rays. The emitted X-rays are called scattered X-rays, and the phenomenon is known as Thomson scattering. The X-rays produced by Thomson scattering are coherent or in phase with the incident X-rays. This means that the scattered X-rays leave the atom in a different direction but with the same wavelength as the incident X-rays. Coherent scattering, also known as elastic scattering, occurs if there is no energy transfer during the collision of the incident X-rays with the atoms.

Inelastic scattering is another type of scattering that occurs when X-rays strike a material. It is associated with the energy loss of the incident beam during the collision of X-rays with a loosely bound electron of the atom. If the energy imparted to the electron exceeds the electron binding energy, the electron is ejected from the atom. A part of the absorbed incident beam energy is converted to the kinetic energy of the electron leaving the atom. This phenomenon is known as photoionisation, and the electron released by the ionisation of an atom is called a photoelectron. The remaining energy of the incident beam is re-emitted as X-rays of lower energy. The wavelength of an incoherently scattered beam is longer than the primary beam owing to energy loss during the collision. An inelastic collision is also called an incoherent collision because of the phase change when the X-ray photons are re-emitted. Another kind of incoherent scattering undergone by incident X-rays is Compton scattering. The X-rays scattered by this phenomenon will have a longer wavelength, i.e., lower energy, because a part of the incident energy is transferred to the electron in the atom during the collision. However, Compton scattering is not accompanied by the ejection of the electron from the atom.

2.5.3 Diffraction of X-rays

The X-rays incident on a specimen get scattered by the atoms present in the crystal lattice of the material. The superposition of the scattered waves produces a resultant wave whose amplitude is the algebraic sum of the amplitudes of the scattered waves. The resultant amplitude may be lower, greater, or the same as that of the scattered waves, depending on the phase difference between the superposed waves. The phase difference, or path difference, of the waves in turn depends on the distance between the source points and the wavelength of the scattered waves. If the waves have a phase difference of even multiples of π, i.e., the waves are in phase, it results in constructive interference of the waves to produce a resultant wave of maximum amplitude. As the path difference increases, the waves become out of phase, and the resultant wave amplitude decreases. A phase difference between the waves in odd multiples of π causes the waves to cancel each other, and the resultant amplitude will be zero. This is known as destructive interference.

The crystal structure of a crystalline or semi-crystalline material consists of a uniformly spaced array of atoms with long-range order. For the X-rays scattered by atoms to undergo diffraction, the interatomic spacing of the crystal should be on the order of the wavelength of the X-rays. As the interatomic spacing of crystals is around 2 Å, the typical X-ray wavelength range chosen for the crystallographic studies is 0.5–2.5 Å. The X-rays coherently scattered by the crystal will superpose to produce constructive interference, and a diffraction pattern is obtained.

2.5.3.1 Laue Equations

Laue developed three simultaneous equations known as Laue equations to describe the diffraction of X-rays by a crystal in three dimensions. According to Laue, all three equations have to be simultaneously satisfied to observe sharp peaks in the diffraction pattern. The Laue equations indicate that maximum diffraction occurs at specific angles for crystal lattices depending on the X-ray wavelength and unit cell dimensions, as given in equations 2.4–2.6:

$$a(\cos\psi_1 - \cos\varphi_1) = h\lambda \quad (2.4)$$

$$b(\cos\psi_2 - \cos\varphi_2) = k\lambda \quad (2.5)$$

$$c(\cos\psi_3 - \cos\varphi_3) = l\lambda \quad (2.6)$$

where λ is the X-ray wavelength, a, b, c are the unit cell dimensions, h, k, l are the Miller indices of the lattice plane, and ψ_{1-3} and φ_{1-3} are incident and diffracted angles formed at parallel lattice planes in three orthogonal directions, respectively.

2.5.3.2 Bragg's Law

Taking into consideration that incident X-rays penetrate deep into a material, Bragg developed a simpler way to describe the diffraction of X-rays from a crystal. He assumed that the diffraction of X-rays from a lattice plane was analogous to the reflection of visible light from a mirror. According to Snell's law for the reflection of light, the angle of incidence of light rays must be equal to the angle of reflection. When a monochromatic beam of X-rays travelling in phase is incident on a crystal at an angle θ, they are scattered by the atoms in the crystal in different directions. A few scattered X-rays make the same angle with the lattice planes as the incident X-rays, i.e., they travel in phase. The superposition of the scattered X-rays that travel in phase undergoes constructive interference and forms a diffracted beam.

Figure 2.4 shows that the X-rays scattered from a plane travel an additional distance of δ compared to the X-rays scattered from a plane lying above it. The extra distance travelled by a wave leads to a difference in phase between the two X-rays. This is known as phase lag or phase shift. The phase lag is observed before the X-rays impinge on the lattice plane and after the X-rays are scattered from the plane. As depicted in Figure 2.4, the X-rays reflecting from the second plane have to travel an extra distance of ABC than the X-rays reflecting from the top plane.

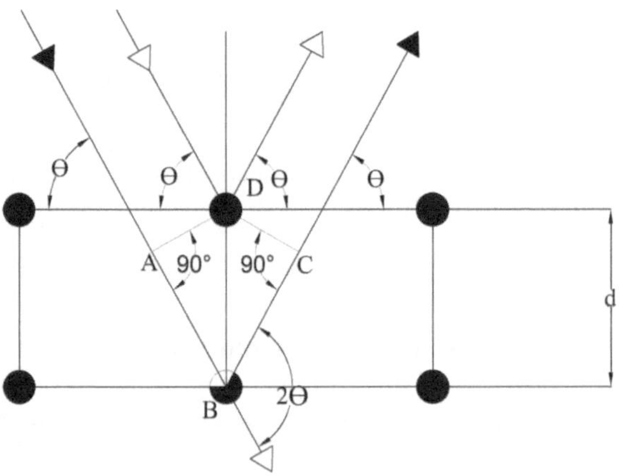

FIGURE 2.4 Illustration of Bragg's law.

For X-rays striking a set of lattice planes at an angle of incidence θ, the phase lag is given by equation 2.7:

$$\delta = AB + BC \tag{2.7}$$

Considering triangles DAB and DCB, AB and BC can be arrived at as shown in equations 2.8 and 2.9:

$$AB = BC = d\sin\theta \tag{2.8}$$

Therefore,

$$\delta = 2d\sin\theta \tag{2.9}$$

where d is the interplanar spacing and θ is Bragg's angle.

The scattered waves should be completely in phase for constructive interference to take place. For two waves to be completely in phase, the phase lag δ should be zero or an integral multiple of wavelength λ, i.e., the distance ABC should be equal to $n\lambda$, where n is an integer (Figure 2.5). Similarly, the phase lag of X-rays scattered from planes lying below should be an integral multiple of λ. Sharp peaks are observed in the diffraction pattern due to constructive interference of scattered X-rays. A point to note is that diffraction of X-rays occurs only when the X-rays are incident on the planes exactly at Bragg's angle. Moreover, constructive interference of scattered X-rays occurs only for crystalline materials, whereas for amorphous materials, diffused scattering occurs. The relation between phase lag and X-ray wavelength is given as shown in equation 2.10:

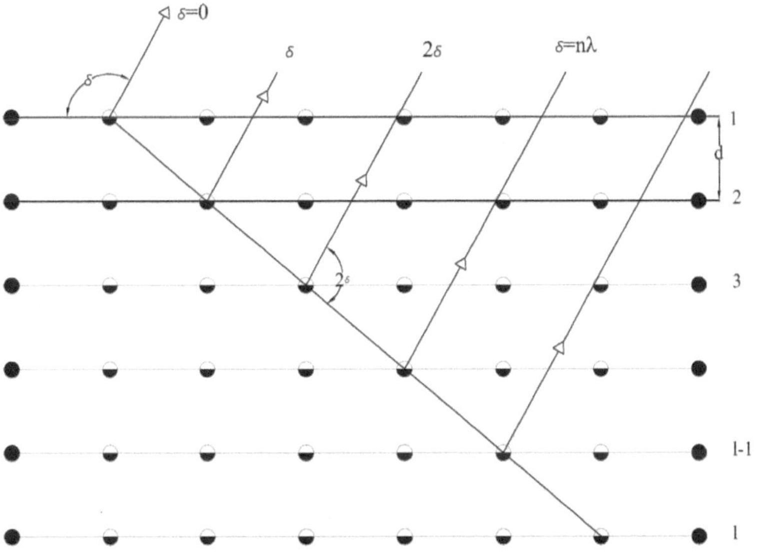

FIGURE 2.5 Phase lag of the scattered X-rays along the depth of crystal lattice.

$$\delta = n\lambda \tag{2.10}$$

Combining equations 2.9 and 2.10 for phase lag, we obtain Bragg's equation, as shown in equation 2.11:

$$n\lambda = 2d\sin\theta \tag{2.11}$$

where n is known as the order of reflection, and for constructive interference, n = 0, ±1, ±2, and so on. n is equal to the path difference between waves scattered by adjacent lattice planes in terms of the number of wavelengths and can take any integer value such that the value of sin θ does not exceed unity. Hence, for a specific wavelength and interplanar spacing, corresponding to n = 1, 2, 3, 4, etc., there are several angles of incidence (θ_1, θ_2, θ_3, θ_4, etc.) at which the X-rays are diffracted from the crystal. In general, diffraction of scattered X-rays occurs in directions in which Bragg's law is satisfied.

For all the other sets of planes where the path difference is not an integral multiple of the wavelength, X-rays undergo destructive interference and diffraction of X-rays does not occur. In such cases, Bragg's law is not satisfied. For example, if the phase shift is 0.5λ for one plane, there will be another plane in the crystal for which the phase shift is λ, thereby resulting in the cancelling out of the scattered X-rays with each other. The scattered X-ray, which is exactly out of phase to cancel out another X-ray, need not be generated from the plane immediately below it. Considering the entire crystal, the particular plane may even lie in another unit cell located at some depth within the crystal.

The possible angles at which diffraction may occur for a cubic system can be determined as follows: if the lattice planes (*hkl*) and wavelength of incident X-rays. The below equation can predict all the incident angles at which diffraction may occur from the planes *hkl* for a particular wavelength. Hence, the diffraction angles can be determined by knowing the lattice parameters and the crystal system for an X-ray of a known wavelength. This means that diffraction directions of incident X-rays are governed solely by the size and shape of the unit cell.

For a cubic system, d is given by equation 2.12:

$$d^2 = a^2/\left(h^2 + k^2 + l^2\right) \tag{2.12}$$

Using equations 2.11 and 2.12, for n = 1, λ is derived as shown in equation 2.13:

$$\lambda^2 = 4\left[a^2/\left(h^2 + k^2 + l^2\right)\right]\sin^2\theta \tag{2.13}$$

However, the actual phenomenon when an X-ray strikes a material is the scattering by the atomic electrons and not a reflection from the lattice planes, as assumed by Bragg. Even though both phenomena have similarities, they are fundamentally different in many aspects. The diffracted beam is formed by scattering of X-rays by all the atoms lying in the incident X-ray path within the material, whereas reflection of light only occurs in the surface layer. Reflection of light can occur at any angle of incidence, whereas diffraction of a monochromatic beam of X-rays occurs only at Bragg's angle. Moreover, the atoms in one plane may not lie exactly below the atoms in the other plane. This is a major drawback in the derivation of Bragg's equation. Therefore, Bloss suggested a slight modification in the assumption. He proved that the equation holds true even if the atom in one lattice plane is shifted by an arbitrary distance in the same plane.

2.6 INFORMATION FROM X-RAY DIFFRACTOGRAM

The XRD pattern can be used to obtain the following information:

i. Identification of crystalline phases in a material.
ii. Distinguish between crystalline and amorphous phases in a material.
iii. Quantitative phase analysis of materials.
iv. Determination of crystal structure.
v. Measurement of lattice parameters.
vi. Indexing of diffraction peaks. Indexing is the process of assigning Miller indices to each diffraction peak. It can be done when all the peaks expected for a crystal structure are present in the diffraction pattern.
vii. Determination of the crystallite size. It is calculated by measuring the peak width in a diffraction pattern. The peak width is defined as the full width at half the maximum height of the peak.

2.7 INSTRUMENTATION

Equipment for powder diffractometric studies mainly consists of three components: an X-ray source, a diffractometer and a detector.

2.7.1 Powder Diffractometer

A powder diffractometer is an instrument used for XRD studies of powder specimens. The Bragg–Brentano arrangement and the Seemann–Bohlin arrangement are the two common types of parafocusing geometric arrangements developed for diffractometers. Most commercially available powder diffractometers use the Bragg–Brentano geometric arrangement illustrated in Figure 2.6. In this arrangement, the X-ray source and the receiving slit are positioned on the circumference, and the specimen is at the centre of a goniometer circle. The goniometer circle is also known as the diffractometer circle. The rotation of these components is controlled by a goniometer, which is the central component of a diffractometer and contains the specimen holder. The X-ray source and the detector are mounted on two arms, and either one or both components can move along the periphery. The specimen holder can be rotated about the axis perpendicular to the plane of the circle. The distance from the specimen holder to the receiving slit is known as the goniometer radius. The goniometer radius of typical diffractometers is in the range of 15–45 cm. The angle between the X-ray source and the specimen surface is called the angle of incidence. They are positioned such that the incident angle is equal to Bragg's angle θ, and the detector

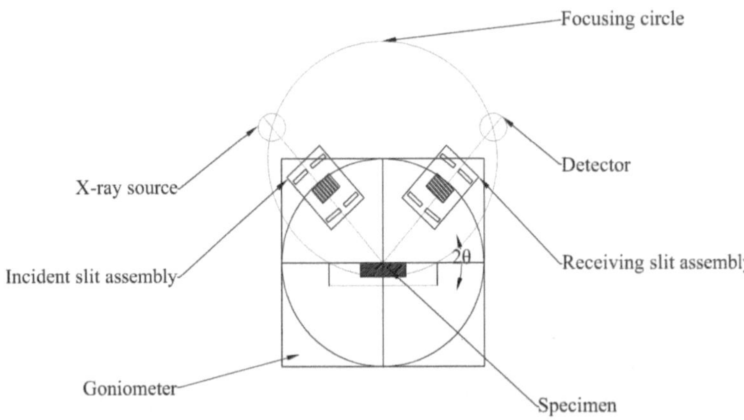

FIGURE 2.6 Schematic representation of the Bragg–Brentano geometry.

is placed at an angle of 2θ to the primary X-ray beam, i.e., the X-ray source. The focusing circle is another circle defined in the Bragg–Brentano arrangement. The X-ray source, the specimen and the detector lie on the circumference of the focusing circle. As the angular position of any of these components is varied, the radius of the focusing circle will change.

The diffractometers are available in vertical and horizontal configurations. The horizontal configuration is preferred when heavy specimens are to be studied. The diffractometers with a vertical configuration are further classified based on the angular movement of their components. In a vertical θ:2θ system, the detector and specimen position are varied such that the incident X-ray strikes the specimen at an angle θ and it makes an angle 2θ with the detector. The position of the X-ray source is fixed in this arrangement. The detector can be moved through a range of 2θ values from 0° to 170°. However, in an experiment, the specimen need not be scanned for the entire angular range, and the scan range is decided based on the crystal structure of the material. In a vertical θ:θ arrangement, the X-ray source and the detector are moved in opposite directions (clockwise and anticlockwise, respectively) in the vertical plane, and the specimen position is fixed. The X-ray source rotates at a rate of θ°/min, and the detector rotates at a rate of $-\theta$°/min. In both cases, the distance between the source and the specimen and between the specimen and the detector is fixed and equal. The θ:2θ system is more common than the θ:θ system due to its mechanical simplicity.

2.7.2 X-Ray Source

An X-ray source for a powder diffraction study has three main components: a line-voltage supply, a high-voltage generator, and an X-ray tube. The typical line-voltage supply is 110 V or 220 V single-phase AC or 220 and 400 V three-phase AC. During periods of heavy demand, variations in the line-voltage supply are observed. If the variation is within ±10%, the high-voltage generator provides a stable voltage supply. A high-frequency generator that produces half-wave, full-wave or constant potential waveforms is most commonly used by manufacturers. The generator has a step-down transformer, and a rectifier mounted inside a high dielectric tank to provide electrical insulation. The function of a transformer is to reduce the line-voltage supply from 220 to 10 V and supply it to the X-ray tube to heat the filament in the tube.

The excitation of the target metal in the X-ray tube occurs only when the critical excitation potential of a characteristic wavelength is overcome by the incident high-energy electrons. Since the voltage supplied to the X-ray tube is higher than the excitation potential only for a specific period of the wave cycle, the emission of characteristic X-rays happens only during this period. This part of the cycle is known as the duty cycle of the generator. The duty cycle is doubled when a full-wave rectified generator is used instead of a half-wave rectified generator. In the case of constant potential generators, the maximum potential is maintained throughout the cycle, and theoretically, the duty cycle of the generator should be 100%. However, practically, some variations in the potential are observed due to the ripple effect.

The two common sources of X-rays are X-ray tubes and synchrotrons. A brief description of the two types is given in the following sections.

2.7.2.1 X-ray Tube

The X-ray tube is a simple and conventional X-ray source used in most laboratories owing to its ease of maintenance. In an X-ray tube, high-energy electrons are made to bombard a target metal, causing the emission of X-rays. The thermal properties of the target metal determine the brightness of the radiation. The rapid deceleration of high-energy electrons results in the conversion of kinetic energy to heat energy. The dissipation of heat also occurs during the impact of the electrons with the target metal. Therefore, continuous cooling of the target metal with chilled water is required to prevent the metal from melting due to heat. In addition to energy loss by heat dissipation, the energy of the X-rays generated is further reduced due to monochromatisation and collimation of the X-ray beam. Hence, the energy of the X-ray beam is significantly lower than the input energy of the high-energy electrons.

i. *Sealed X-ray tube:*

   A sealed X-ray tube is an assembly of a stationary anode coupled with a cathode placed inside a sealed chamber, as shown in Figure 2.7. The X-ray tube also has beryllium windows and a cooling system. High The chamber is made of metal, glass or ceramic material. It is kept under a high vacuum to prevent the collision of air particles with the primary electrons and generated X-rays. Copper is generally used as the anode for powder diffractometry and molybdenum for single-crystal diffractometry. The lower the atomic number (Z) of the target metals, the longer the wavelength of $K\alpha$ radiation. In decreasing order of atomic number, silver, molybdenum, nickel, cobalt, iron, and chromium are other metals that can replace copper as the target metal. Metals with a lower Z than chromium produce radiation with longer wavelengths, which are heavily absorbed by the tube window. On the other hand, metals with a higher Z than silver generate shorter-wavelength radiation with dominant background white radiation, thereby making such metals less useful for crystallographic study. Molybdenum is mainly preferred

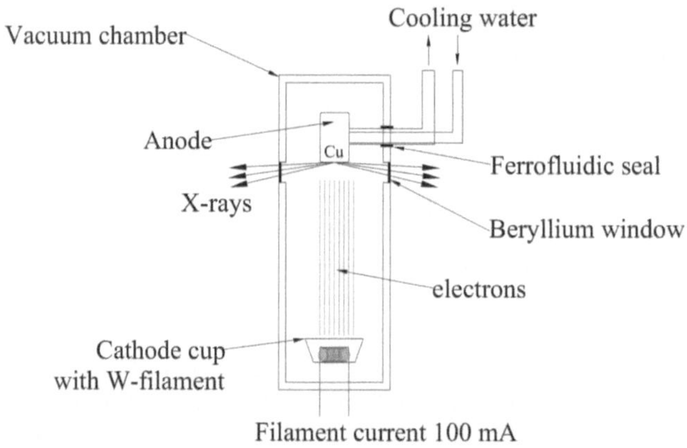

FIGURE 2.7 Sealed X-ray tube.

for samples with small unit cells due to its short wavelength and when the sample has a higher absorption of Cu *Kα* radiation. However, the use of molybdenum as a target metal has a limitation. Since most high-voltage generators have a maximum operating voltage of 60 kV, it is challenging to achieve the optimum voltage required for the excitation of Mo *Kα* radiation (80 kV). Cr *Kα* (2.2896 Å) has a comparatively longer wavelength than Cu *Kα* (1.5405 Å). Hence, chromium anodes are useful for developing diffraction patterns of materials with unit cells of larger interplanar distance, such as organic matter. However, chromium has lower thermal conductivity than copper, and the air in the tube absorbs 50% of the Cr *Kα* radiation compared to 10% absorption for Cu *Kα* radiation. Iron is preferred over copper as a target metal for the crystallographic study of materials rich in iron. This is because impinging the materials rich in Fe with Cu *Kα* produces X-ray fluorescence of Fe *Kα*, and the signal-to-noise ratio is very low.

A standard tube has two pairs of opposite windows, each located at 90° spacing around the circumference of the tube. One pair of opposite windows is used to produce a point-focused beam, whereas the other pair is used to obtain a line-focused beam. The point-focused beam is useful for single-crystal diffraction owing to the smaller size of a single crystal. On the other hand, powder diffraction requires irradiation of a higher number of particles in the specimen and, hence, employs a line-focused beam. Even though the intensity of both beams is nearly the same, a point-focused beam is brighter than a line-focused beam. The windows used in the X-ray tube are made of beryllium, which is vacuum-tight but transparent to X-rays.

The tungsten filament in the sealed tube is electrically heated to emit electrons. The anode and the metallic body are always grounded, and the cathode is maintained at a high negative potential to develop a potential difference between the cathode and the anode. Surrounding the filament is a Wehnelt cap, which is at a slightly higher negative potential to ensure fine focusing of the electron beam into a small focal spot on the target metal.

The electrostatic potential difference (30–60 kV) created between the anode and cathode causes the electrons to accelerate towards the target metal. The bombardment of the target metal with high-energy electrons leads to the generation of X-rays. The generated X-rays leave the X-ray tube through the beryllium windows provided in the cylindrical body of the X-ray tube.

The efficiency of sealed X-ray tubes is very low (less than 1%) because a significant part of the kinetic energy of electrons striking the anode is converted to heat. Therefore, continuous cooling of the anode is recommended to enhance the conversion of electrons to X-rays after the collision. The contamination of the anode surface due to the deposition of tungsten from the cathode is another limitation of the sealed X-ray tube. Pitting of the anode is also reported to reduce the lifetime of the X-ray tube. Hence, demountable fixed anode tubes are preferred so that they can be periodically demounted and cleaned.

ii. *Rotating anode X-ray source:*

One solution to the problem of the low efficiency of the conventional X-ray tube is to use a rotating anode tube, as shown in Figure 2.8. In this tube, the continuous rotation of the anode brings fresh target metal into the focal spot area. The rotating anode is cooled using chilled water to avoid overheating the target metal. The continuous rotation ensures that the cooler part of the anode is always available for electron bombardment. Moreover, the irradiation area of the anode is significantly increased, implying that energy dissipation is spread over a larger area. Both of these factors result in higher thermal efficiency. The rotating anode as an X-ray source enhanced the efficiency by up to 20 times higher than standard sealed X-ray tubes. The efficient cooling of the anode increases the brightness of X-rays generated and results in better diffraction patterns.

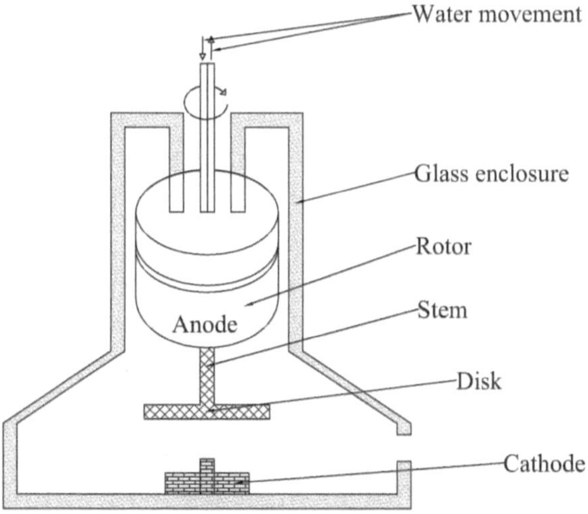

FIGURE 2.8 Rotating anode assembly.

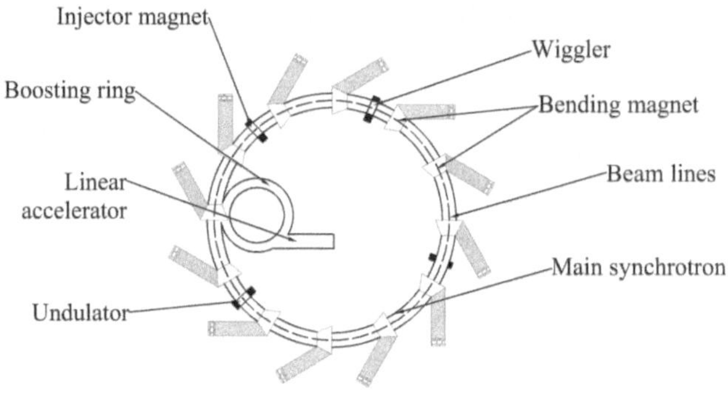

FIGURE 2.9 Synchrotron.

iii. *Synchrotron:*

A synchrotron is the most advanced and powerful X-ray source (Figure 2.9). Electrons are periodically injected from a linear accelerator into a booster synchrotron. The function of a booster synchrotron is to enhance the energy of the injected electrons. Dipole-bending magnets control the motion of the accelerated electrons, and the electrons start to follow a polygonal path. The high-energy electrons are transmitted to a storage ring or main synchrotron, and their motion is confined inside the ring. Each time the electrons pass through the bending magnets, they are accelerated. This results in energy loss in the form of EM radiation ranging from radio waves to X-rays. The X-ray beam is ejected from the synchrotron in a direction tangential to the electron orbit. The X-rays emitted from a synchrotron have exceptionally high brightness (10^4–10^{12} times higher) and are highly polarised compared to conventional X-ray tubes. Another advantage of using synchrotrons is the availability of a highly intense X-ray beam over a broad range of wavelengths, which leads to an easier selection of wavelength desired for the experiment. However, the construction and maintenance costs of synchrotrons are high, making them less popular.

2.7.3 Slit Assembly

A receiving slit assembly consisting of an anti-scatter slit, Soller slits and a receiving slit is placed in the path of the diffracted X-rays from the specimen. The anti-scatter slit allows X-rays only from the specimen area to pass through it. Thus, the anti-scatter slit helps to reduce background radiation due to scattering from the air and the specimen holder. After the anti-scatter slit, the diffracted beam passes through Soller slits and a receiving slit before reaching the detector. The receiving slit reduces the divergence of the diffracted beam, and the width of receiving slit controls the X-ray intensity. The larger the slit width, the higher the peak maximum in the diffraction pattern. A diffracted beam monochromator may be placed between the receiving slit and the detector to remove unwanted wavelengths and reduce the background radiation in the diffracted X-ray beam.

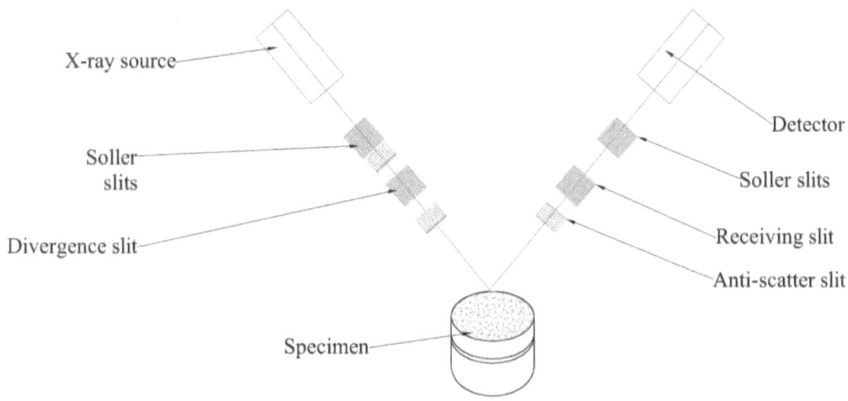

FIGURE 2.10 Arrangement of slits in a diffractometer.

Figure 2.10 shows the arrangement of slits in an X-ray diffractometer. The incident slit assembly is located in the path of the primary X-ray beam. It consists of Soller slits and divergence slits. Soller slits are a set of equally spaced, parallel, thin plates that reduce the axial divergence of the primary X-ray beam. Commonly, the plates are made of metals with a high atomic number, such as molybdenum, owing to their high absorption capacity. In addition, divergence slits are kept between the X-ray source and the specimen to obtain a collimated beam of narrow width. The function of a divergence slit is to restrict the vertical divergence of the beam. It helps to avoid irradiation of the specimen holder by the X-rays.

2.7.4 Specimen Holder

The common types of specimen holders used in an XRD study are presented in Figure 2.11. The holders are made of metal, plastic or glass materials. They mostly have a cylindrical or rectangular, shallow cavity in which the specimen is loaded. The cavity is designed taking into consideration the half-depth of X-ray penetration

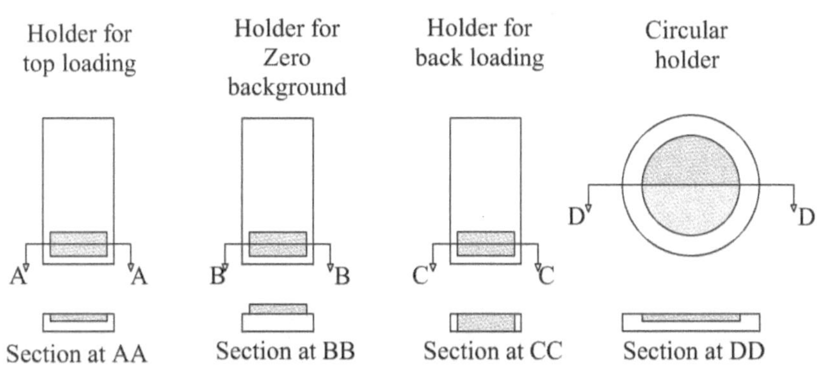

FIGURE 2.11 Specimen holders.

in the specimen. It is a critical factor since it determines the effective diffracting surface and, therefore, the peak position in the diffraction pattern. In certain holders, instead of providing a cavity, a spot with a rough surface is created over which a thin layer of powder can be spread. The rough surface is prepared by sandblasting with hard particles.

2.7.5 Detectors

An X-ray detector is used to measure the intensity of the diffracted X-rays in a diffractometer. It mainly consists of three components: a transducer, a pulse formation circuit and a counting circuit. The X-rays collected by the detector are converted into electric current using the transducer. Further, the electric current is converted to electric pulses using a pulse formation circuit. The relatively low-voltage pulses generated are amplified by a linear amplifier and sent to a pulse shaper, giving square-shaped pulses as output. Finally, the counting circuit measures the number of electric pulses per unit time. The number of electric pulses is directly proportional to the X-ray intensity entering the detector.

2.7.5.1 Desired Characteristics of a Detector

i. *Quantum-counting efficiency:*

   The efficiency of a detector to collect the diffracted X-rays incident on it is known as quantum-counting efficiency. A detector with high quantum-counting efficiency effectively collects diffracted X-rays of the desired wavelength and discriminates undesirable short-wavelength radiations, such as bremsstrahlung X-rays.

ii. *Linearity:*

   A detector is considered to be linear when the rate of incidence of X-rays on the detector is directly proportional to the rate of voltage pulses generated by the detector. Usually, a short duration known as the dead time of the detector is required to collect and process an X-ray and then reset the detector for the next operation. During this time, the detector will be inactive, and there will be a slight lag in the collection and processing of the subsequent X-ray. At high photon flux, an X-ray might enter through the detector window while the previous X-ray is still under processing, i.e., during its dead time. In such a situation, a small fraction of the X-rays striking the detector may not be converted into voltage pulses, and they are not recorded. A detector incapable of collecting and counting every X-ray entering the detector is said to be non-linear. The loss of linearity and dead time of a detector must be as minimal as possible.

iii. *Energy proportionality:*

   The energy of an X-ray striking the detector determines the current produced by the transducer and, thereby, the number of ionising events resulting in the generation of voltage pulses. Hence, the output signal of a detector is related to the energy of incident X-rays. A detector is considered proportional if the amplitude of the generated voltage pulse is proportional to the incident X-ray energy. The pulses with too high and too low amplitude correspond to X-rays of very short and very long wavelengths, respectively. A high proportionality detector can be used to produce a monochromatic beam of X-rays by not counting the low- and high-amplitude pulses.

iv. *Energy resolution:*

The ability of a detector to distinguish two X-rays of different wavelengths is known as the resolution of the detector. Each of the output signals from a detector has a spread that is related to the resolution. The energy resolution can be calculated as shown in equation 2.14:

$$R(\%) = \frac{\sqrt{\Delta V}}{V} \times 100 \quad (2.14)$$

where V is the maximum voltage in the pulse distribution and ΔV is the peak width at half-maximum of the pulse amplitude distribution. ΔV is also known as the spread of voltage pulses. Narrow peaks in the diffraction pattern indicate high resolution. Better resolution leads to low background radiation, a high signal-to-noise ratio and better detection limits.

The different types of detectors used in a conventional powder diffractometer are gas proportional detectors, scintillation detectors and semiconductor detectors. The various components and working principles of these detectors are described below.

2.7.5.2 Gas Proportional Detectors

A gas proportional detector consists of a grounded cylindrical shell (cathode) and a thin wire placed along the central axis (anode) maintained at a positive potential (Figure 2.12). A mixture of a noble gas (argon, xenon or krypton) and a quenching gas (chlorine, CO_2 or CH_4) is used as the filling gas inside the cylinder. A small window is provided at the side of the cylinder through which the X-rays enter the detector. Mica or beryllium windows are most popular owing to their high transparency to X-rays of longer wavelengths. Another window is provided on the opposite side for the unabsorbed X-rays to exit the chamber.

The X-rays entering the gas chamber interact with noble gas atoms present in the chamber. The absorption of X-rays leads to the release of photoelectrons and leaves the atoms in an ionised state. An electron and a positively charged ion combination are known as ion pairs. The number of ion pairs discharged depends on the incident X-ray energy and the binding energy of the atomic electron. The free electrons accelerate towards the anode owing to the large potential difference between the anode and the cathode. At the same time, the positive ions move in the opposite direction towards the casing, which is at zero potential. If the potential difference is sufficiently large, the electrons produced are rapidly accelerated towards the anode. They gain enough kinetic energy to knock out electrons from other noble gas atoms. The collision leads to further ionisation of the atoms and the release of secondary electrons. The cumulative discharge of ion pairs is known as gas amplification or avalanche production. The current resulting from the movement of ion pairs is amplified and recorded. The number of current pulses measured is proportional to the energy of X-rays absorbed by the gas atoms. Hence, this type of detector is known as

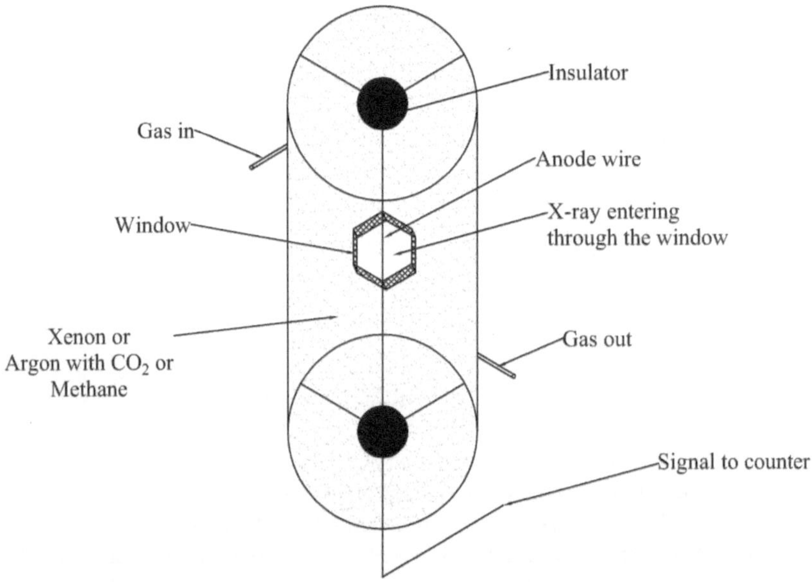

FIGURE 2.12 Gas proportional detector.

a proportional detector. The proportionality of a detector is an important factor as it helps to distinguish X-rays of different wavelengths.

Gas proportional detectors have very high quantum-counting efficiency. It can be attributed to the low absorption of short-wavelength radiation by the gas present in the detectors. Moreover, owing to the high resolution and energy proportionality of these detectors, the pulse heights can be analysed and discriminated to sort out X-rays of different wavelengths entering a detector. Pulse height discrimination allows the detector to remove current pulses corresponding to short- and long-wavelength X-rays and select only X-rays of the desired wavelength for counting. β-filters are used in combination with pulse height discrimination to eliminate $K\beta$ and white radiation with higher efficiency.

2.7.5.3 Scintillation Detectors

A crystal scintillator and a photomultiplier tube are the main components of a scintillation detector (Figure 2.13). The conversion of X-ray energy into electric pulses is a two-stage process. Firstly, the scintillator absorbs the incident X-rays of a certain wavelength and then re-emits the radiation as light photons. The X-rays of shorter wavelengths are prevented from entering the detector. In the second stage, the photomultiplier tube converts the light energy into electric pulses.

The X-ray photons enter the detector through a beryllium window and strike the atoms of the scintillator. Electrons are ejected from the atom due to the absorption of incident energy, resulting in electron jumps from higher to lower energy states and emission of fluorescent photons in the visible spectrum. A flash of light known as scintillation is produced for every X-ray photon absorbed. The amount of fluorescent

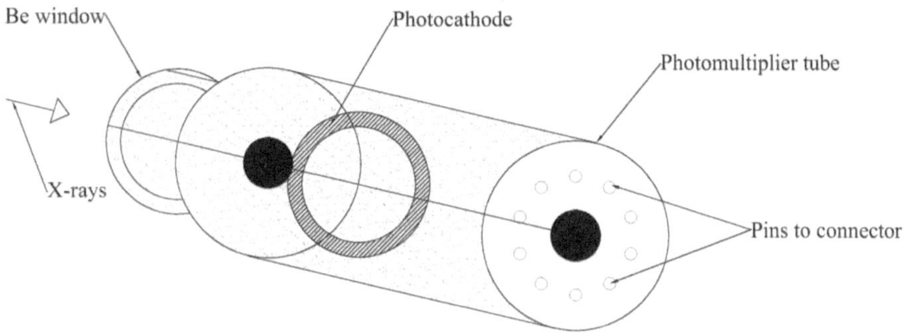

FIGURE 2.13 Scintillation detector.

photons emitted is proportional to the energy of X-rays absorbed. These light photons strike a photocathode in the photomultiplier tube. A burst of electrons is released and accelerated towards a chain of photomultiplier electrodes, known as dynodes. When the electrons strike the first dynode, secondary electrons are emitted that are then accelerated towards a second dynode. At the subsequent dynodes, the same process of absorption and re-emission of electrons is repeated. Each dynode is maintained at a higher positive potential compared to the preceding dynode. Hence, the kinetic energy of electrons leaving a dynode is successively higher and is utilised to release more electrons from the subsequent dynode. An anode collects the electrons produced by the last dynode, and a voltage pulse is generated. The output pulse is then fed to an amplifier and the counting circuits. Even though the energy resolution of a scintillation detector is low due to a large number of losses, it has high quantum efficiency. Similar to the gas proportional detector, the pulse size is proportional to the energy of X-ray photons absorbed by the scintillation detector.

2.7.5.4 Semiconductor Detectors

A lithium-drifted silicon [Si(Li)] detector is the most commonly used semiconductor or solid-state detector. It is a crystal made of intrinsic silicon sandwiched between a *p*-type and an *n*-type layer, forming a *p–i–n* diode. A thin layer of gold is deposited on the two faces of the crystal for electrical contact. When X-rays strike a Si(Li) detector, the electrons in the valence band are excited and jump to the conduction band. It creates a hole or vacancy in the valence band. The electron-hole pairs generated are directly proportional to the incident X-ray energy. A reverse-bias potential is applied to the crystal, causing the electrons and holes to accelerate towards oppositely charged electrodes. The movement of electron-hole pairs leads to the generation of electric current pulses. The pulse amplitude is directly related to the number of electron-hole pairs created and, therefore, to the energy of incident X-rays (Figure 2.14).

These detectors must be kept at liquid nitrogen temperature (77 K) at all times to minimise noise generated due to the thermal excitation of electrons at room temperature. Maintaining the temperature of the detector at 77 K also helps to minimise the thermal diffusion of Li, which may alter the diode properties. The Si(Li) detector is more efficient than a gas proportional detector for pulse height analysis owing to its

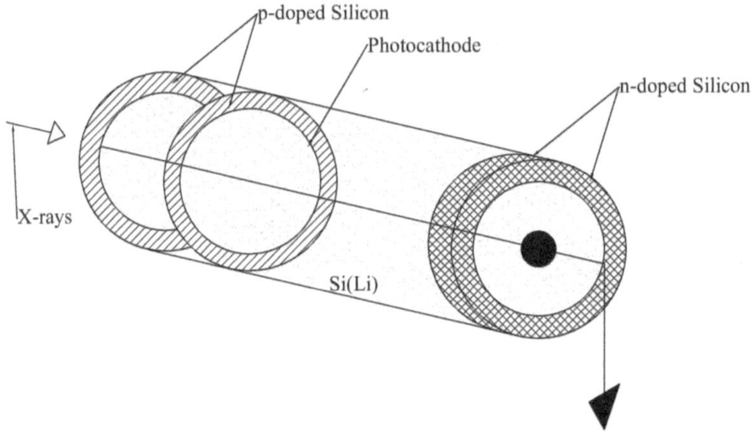

FIGURE 2.14 Semiconductor detector.

higher energy proportionality and superior resolving power. The shorter wavelengths are prevented from striking the diode by choosing an optimum thickness of the silicon layer.

An intrinsic germanium detector is another type of solid-state detector. The main advantage of using a germanium detector is the higher absorption of X-rays by germanium compared to silicon. Moreover, the energy required to form an electron-hole pair is only 2.9 eV for Ge, whereas it is 3.8 eV for Si. Therefore, more electron-hole pairs are generated by an intrinsic Ge detector than a Si(Li) detector when X-rays of the same energy are incident on the detector. It gives rise to better resolution and a higher signal-to-noise ratio for an intrinsic Ge detector. However, the Si(Li) detector is more popular owing to its higher efficiency, higher reliability, and ease of manufacturing. In addition to the high quantum-counting efficiency of the Si(Li) detector, it has the ability to collect X-rays of a particular wavelength selectively. Therefore, the detector can be used in place of monochromators to eliminate $K\beta$ radiation.

The X-ray detectors are also classified into point, line and area detectors based on whether, in addition to recording the intensity of diffracted X-rays, the detector can resolve the location where the photon has been absorbed. In the case of point detectors, spatial resolution is not possible as they record the intensity of the diffracted beam at one point at a time. Conventional gas proportional detectors, scintillation detectors and semiconductor detectors are examples of point detectors. Line and area detectors are other types of detectors in which spatial resolution in one direction and two directions, respectively, is possible. A position-sensitive detector is an example of a line detector, and charge-coupled devices and image plates are examples of area detectors.

2.7.5.5 Counting Circuits

Two kinds of circuits are available for measuring pulse rates: a counting rate meter and a scaler/timer.

i. *Counting rate meter:*

A counting rate meter continuously monitors the current or voltage signal and records the pulse rates in counts per second (cps). The series of electric pulses is converted to a steady current to obtain continuous readings of X-ray intensity. At the start of scanning, the detector position is set at $2\theta = 0°$. The detector is connected to the counting rate meter, and the output of the circuit is fed to a strip chart recorder. During the scanning, the detector is moved through increasing 2θ values at a constant angular velocity until the complete scan range is covered. Simultaneously, the chart paper in the recorder is moved at a constant speed such that the distance along the chart length is proportional to 2θ. The chart shows a plot of diffracted X-ray intensity (cps) versus diffraction angle 2θ. A calibrated voltmeter can also be used to display the recorded data. Measurement using a counting rate meter gives a continuous diffraction pattern.

ii. Scaler/timer

A scaler is an electronic device connected to a detector used for counting the electric pulses generated by a detector. During the scanning, the angular position of the detector is set at a particular value of 2θ for a preset duration. The scaler counts the number of electric pulses for that duration. The timer sends a stop pulse when the preset time is reached. After counting the pulses, the detector is rotated to a new angular position, and the pulse count is again recorded. In a similar manner, the detector is rotated at fixed angular increments until the complete scan range is covered. The plot between diffracted intensity and diffraction angle is manually plotted. However, continuous readings of X-ray intensity cannot be measured by this method.

2.8 WORKING PRINCIPLES

An X-ray beam emerging from the X-ray source is divergent in nature and is passed through a slit assembly consisting of Soller slits and divergence slits to produce a narrow collimated beam. The collimated X-ray beam is focused on the specimen surface at an angle θ and is then scattered by the specimen at an angle 2θ with respect to the incident X-ray beam. A receiver slit assembly (an anti-scatter slit, Soller slits and a receiving slit) and a monochromator are placed in the path of the diffracted beam for collimation and monochromatisation of the diffracted beam. The diffracted beam is then collected by the detector, whose angular position is measured as the diffraction angle 2θ. For the θ:2θ diffractometer arrangement, the supports of the detector and the specimen stage are mechanically coupled such that during the scanning, an angular variation of the detector position by 2θ degrees is automatically accompanied by a variation of the incident angle (rotation of the specimen) by θ degrees. The detector records the intensity of scattered X-rays in counts or cps at each 2θ angle. The acquired raw diffraction data is processed to obtain a diffraction pattern. Data smoothing, background reduction, stripping of $K\alpha_2$ radiation, and peak search are the common data processing methods adopted. The diffraction pattern is stored as a list of the diffracted intensities and Bragg's angles.

2.9 DIFFRACTION PATTERN

A plot of the intensity of the diffracted X-rays as a function of the diffraction angle 2θ or interplanar spacing is known as the XRD pattern. The peak intensity gives a measure of the number of X-rays of a specific wavelength recorded for each 2θ angle. For small values of 2θ, a single sharp peak is observed, whereas for larger values, a pair of peaks corresponding to $K\alpha$ doublet may be detected. Sometimes, the scale of the 2θ axis can be increased to observe the $K\alpha_1$ and $K\alpha_2$ lines separately, even at lower 2θ values.

2.9.1 Single-Crystal Diffraction Pattern

In single-crystal diffraction, a particular set of lattice planes is oriented towards the incident X-rays. The diffraction of X-rays striking the crystal occurs at Bragg's angle, which is determined using Bragg's equation. The crystal is oriented in different directions during the experiment, and the detector scans and collects the diffracted X-rays. For each orientation of a single crystal, a different set of lattice planes are in the position to diffract the incident X-rays. Since the diffraction pattern varies with the *d*-spacing of the lattice planes, the orientation of the crystal is critical.

2.9.2 Powder Diffraction Pattern

Polycrystalline materials, or powders, are composed of a large number of randomly oriented crystallites. The crystallites are arranged in multiple directions, due to which different lattice planes, each having a unique *d*-spacing, are in a position to diffract the incident X-ray beam. According to Bragg's law, the incident collimated beam of X-rays is diffracted by various lattice planes of the crystal structure at unique diffraction angles depending on the interplanar spacing. Several diffraction peaks are detected in the diffraction pattern, and each peak is characteristic of a specific set of lattice planes.

The intensity of peaks depends on the incident X-ray energy, the depth of X-ray penetration, the nature of atoms and the arrangement of atoms in a unit cell. For an X-ray of a certain wavelength, the position of peaks is governed by the interplanar spacing of lattice planes. Since the atomic structure is unique for a particular material, the XRD pattern provides information about the crystal structure of that material. To obtain diffraction peaks at precise positions, the alignment of the diffractometer should be correct, and the specimen should be appropriately mounted on the specimen holder. Moreover, X-rays must penetrate deep into the specimen so that the maximum number of crystallites contributes to the diffraction data for each phase in the material. This is to ensure that the peak intensities obtained are accurate. Unlike well-defined sharp peaks detected for a crystalline material, the XRD of amorphous materials shows a broad, poorly defined hump that extends over several 2θ angles. The absence of a periodic array of atoms with long-range order is the reason for the poorly developed diffraction pattern of amorphous materials.

2.9.3 Background Radiation

The sharp, intense peaks in the diffractogram are also superimposed by undesired peaks of minimum intensity or background radiation. The background radiation is mainly due to diffused scattering (coherent and incoherent scattering) from the specimen and air and scattering of incident X-rays by the specimen holder, specifically at low values of 2θ. The incoherent diffuse scattering is more intense for specimens with lower atomic numbers. On the other hand, coherent diffused scattering is due to imperfections in the crystals. Temperature-diffused scattering is another form of coherent scattering that occurs in soft materials with a low melting point. The diffractometer is maintained under a vacuum or filled with light gases such as hydrogen or helium to avoid diffused scattering from the air.

In addition to the scattering of incident X-rays, fluorescent radiation may be produced from a specimen when the incident X-ray energy overcomes the excitation potential of the bound electrons in an atom. If the fluorescent radiation falls within the acceptable range of the monochromator or the detector, it is collected by the detector. The fluorescent radiation processed by the detector will appear as a background in the diffraction pattern.

Moreover, diffraction of the continuous radiation component of the incident X-rays may form a low-intensity background in the XRD pattern. The diffractogram may also contain undesirable peaks due to contamination from the tungsten filament or beryllium window in the X-ray tube. The intensity of these weak peaks increases as the age of the X-ray tube increases. However, using a monochromator or a β filter can significantly reduce the intensity of the undesirable radiation.

2.10 INFLUENCING FACTORS

There are several specimen properties and instrument parameters that influence the quality of XRD data. A few of the important factors are described in the following sections.

2.10.1 Specimen Characteristics

i. *Particle orientation:*

   To obtain an accurate diffraction pattern for a powder or polycrystalline sample, the most important requirement is to have a completely random orientation of crystallites. The incoming X-rays are incident on a large number of crystallites oriented in different directions with respect to each other. The lattice planes corresponding to each of these crystallites have a different *d*-spacing, and each plane diffracts the incident X-rays at a specific 2θ angle. The diffraction pattern of a polycrystalline sample will have several peaks corresponding to different *d*-spacing values. If the particles tend to have a preferred direction of orientation, the specimen is said to suffer from preferred orientation effects. The absence of complete randomness in the particles in the specimen will affect the precision of the diffraction data.

ii. *Particle shape:*

The particles in a specimen are oriented in the most random and ideal fashion when they are spherical or nearly spherical in shape. It is difficult to achieve the same randomness for most of the other particle shapes. The larger the difference between the longest and shortest dimensions of a particle, the more difficult it is for the particles to orient themselves in random directions. In such cases, the specimen suffers from preferred orientation effects. Usually, the grinding of samples leads to the formation of anisotropic particles that have non-random orientations. Particles with platelet-like or needle-like shapes are commonly found in ground samples. The preferred orientation of platelet-like particles is unidirectional (along the axis of their largest face), as they tend to align with their largest faces parallel to each other. However, the orientation of needle-shaped particles is comparatively more random than that of platelet-shaped particles. Hence, special care should be taken while grinding the sample and mounting the ground samples on the specimen holders.

iii. *Particle statistics:*

The number of crystallites present in the powder that diffract the incident X-rays at a certain angle is called particle statistics. It should be very large or infinite in number to obtain a representative diffraction pattern. Particle statistics errors result in patterns with a few unexpected sharp peaks and other peaks that may be too low. The change in peak shapes leads to problems in profile fitting and quantitative phase analysis. Specimen rotation during scanning enables more particles to diffract at a certain angle and reduces particle statistics errors. Powders composed of particles of a small mean size should be prepared by grinding to minimise the error. However, excessive grinding should be avoided as it may cause structural damage to the sample.

iv. *Particle size:*

The samples have to be appropriately ground to obtain crystallites of sizes finer than 10 μm. The diffraction pattern of coarse-grained samples shows peaks of unusual intensity due to sharp reflections from the poorly ground crystals. On the other hand, excessive grinding of soft materials reduces the crystallinity of the sample and causes the broadening of peaks. The peak broadening effect occurs for crystallites of small size due to the absence of planes that may otherwise cancel out the X-rays, which are not scattered at Bragg's angles. Hence, the peak in the diffraction pattern shows intensity at a range of angles around Bragg's angle, resulting in a broader peak instead of a single sharp peak at Bragg's angle. The peak broadening increases as the size of the crystallite reduces. A heterogeneous material should be carefully ground for an optimum duration such that the harder particles become finer and the crystallinity of the softer particles is not reduced.

v. *Specimen thickness:*

The specimen thickness should be such that all the incident X-rays interact with the atoms of the specimen and are not simply transmitted through the specimen. The intensity of diffracted X-rays is proportional to the irradiation volume of the specimen. Hence, the specimen should have a sufficient thickness so that the diffracted beam intensity is not too low.

vi. *Specimen length:*

The projection of an incident X-ray beam on a specimen at any incident angle should not exceed the size of the specimen. If the spot size of the X-ray beam exceeds the specimen size, then the measured intensity of the X-rays will be underestimated since only a fraction of the beam illuminates the specimen surface. It must be ensured that the length of a flat specimen is large enough so that the irradiated length of the X-ray beam is less than the length of the specimen. Typically, the length of the specimen is chosen to be below 25 mm. The issue of a beam fraction falling outside the specimen area is more common at a low angle of incidence. The corresponding intensity of the diffracted beam is also relatively lower compared to the intensity of X-rays diffracted at higher angles. Moreover, the X-rays falling outside the specimen area may be scattered by the sample holder to produce unwanted peaks or background intensity in the diffraction pattern. In most cases, the divergence slit width is reduced to decrease the size of the incident beam so that the X-ray beam falls entirely within the specimen area.

vii. *Specimen uniformity:*

The penetration depth of an X-ray beam and the interaction volume of the specimen are functions of the incident angle and the packing density of the specimen. A non-uniform packing density leads to random variation in the number of particles in the interaction volume and consequent changes in the intensity of the diffracted beam. Even though compacting the powder filled in the mould will ensure uniform packing, it is not a suitable option since it increases the preferred orientation effects of the powder. Another method to improve uniformity is to rotate the sample rapidly during data collection. Spinning the sample during scanning results in the availability of more crystallites in a position to diffract the incident X-rays.

2.10.2 Instrument Parameters

i. *Detector dead time:*

When a detector used in a diffractometer is not linear, the count rate of pulses recorded by the detector is not directly proportional to the rate of X-ray photons entering the detector. The time lag in the collection and processing of X-rays by a detector is known as the dead time of the detector. All the peaks in a diffraction pattern are normalised with respect to the strongest peak, and since the dead time loss is higher for stronger peaks, a longer dead time may increase the intensity of weaker peaks. Hence, the loss of linearity and dead time of a detector must be as minimal as possible.

ii. *Divergence slit aperture:*

At a specific angle of incidence (orientation of specimen plane), the vertical divergence of the X-ray beam and the irradiation area of a specimen surface depend on the divergence slit width. A narrow divergence slit produces sharper diffraction peaks at a lower intensity and a higher resolution. If the divergence slit width is considerably large, the X-rays may get scattered by the specimen holder at low incident angles, resulting in background radiation. On the other hand, if the slit width is very small, the complete specimen area may not be illuminated by the X-ray beam. Variable divergence slits are also employed for certain diffractometers, using which the angle of view subtended by the divergence slit is varied as the diffraction angle changes. It leads to a constant specimen irradiation area. Generally, most diffractometers use fixed divergence slits due to the higher accuracy of the X-ray intensity measurements.

iii. *Receiving slit aperture:*

The aperture of a receiving slit significantly influences the peak intensity and peak shape in a diffraction pattern. If the diffracted beam width at the slit is nearly the same as the receiving slit aperture, all the diffracted X-rays will pass through the slit. The peak intensity and resolution are not affected in this case. However, using a narrower slit causes a lowering of the peak intensity with less effect on resolution. Conversely, wider slits will distort the peak shape and result in poor resolution. The distortion of shape is because even the non-focused X-rays will pass through a slit of larger width. Hence, there is an optimum size of receiving slit for an experimental setup.

2.10.3 Operating Conditions

The selection of suitable operating conditions is equally important as proper specimen preparation to obtain good-quality diffraction data. The operating parameters of an XRD study include instrument parameters such as X-ray wavelength, monochromatisation, slit apertures, power settings, and data collection parameters such as scan mode, scan range, step size, and counting time. The main factors affecting the operation of an X-ray tube are the power rating of the tube, the type of generator used, the takeoff angle of the tube, and the desired lifetime of the tube.

i. *Operating current and voltage:*

The intensity of X-ray photons generated in an X-ray tube mainly depends on the voltage and current supplied to the tube. Selecting the optimum operating condition of the X-ray tube is essential to maximise the photon flux. The accelerating voltage developed in the X-ray tube influences the type of X-rays produced. If it is significantly low, bremsstrahlung X-rays alone will be present in the spectrum. As the accelerating voltage increases, the intensity of all the wavelengths in the X-ray continuum increases, and a shift in the short wavelength limit towards the left of the X-ray spectrum is observed. On further increasing the voltage, the electrons gain sufficient energy to knock off the inner shell electrons from the target metal, resulting in the release of X-rays. This gives rise to the appearance of sharp peaks corresponding to the characteristic radiation at specific wavelengths, superimposing the white radiation in the X-ray spectrum. Hence, the accelerating voltage supplied should be such that the incident electron energy overcomes the excitation potential of the target metal atoms. As the excitation potential for each metal is different, the operating voltage has to be chosen accordingly. The typical operating conditions when copper is used as the target metal are 40 kV and 30 mA. The optimum voltage and current for a specific takeoff angle are chosen from an isowatt curve, which is a plot between the X-ray intensity from the tube and the operating voltage at various takeoff angles.

The maximum power rating of the X-ray tube is another parameter that should be considered while selecting the operating current and voltage. The mathematical product of filament current and voltage is known as the power rating of the tube. Both parameters should be chosen such that the power rating does not exceed the maximum rating of the tube. The heat dissipation capacity of the anode governs the maximum power rating. The optimum current and voltage selected must also fall within the acceptable limits for operation as per the power curve. The power curve of an X-ray tube is a plot of current versus voltage. Too high a current at a low potential is not acceptable since it shortens filament life due to excessive heating.

ii. *Wavelength of X-rays:*

The X-ray wavelength depends on the target metal used in the X-ray tube to generate X-rays. The typical target metals are copper, chromium, iron, cobalt, nickel and molybdenum. X-rays with longer wavelengths are produced when metals with lower atomic numbers are used. Copper $K\alpha$ radiations are the most preferred X-rays since the dispersion of the diffraction pattern is minimal. Moreover, the peak intensities of X-rays diffracted by most common materials lie within a reasonable scan range (2θ). When radiation of a wavelength longer than copper $K\alpha$ is used, the dispersion of the pattern is increased, and a shift in the pattern to a higher angular range is observed. It reduces the recording range to smaller 2θ values.

iii. *Angle of the specimen holder:*

Most of the commercially available diffractometers use fixed divergence slits, and the operator has to choose a slit of the required width from several slits supplied with the instrument. If the divergence slit width is fixed, the angle of the specimen governs the irradiation area of the specimen surface. The incident angle of X-rays from the source is varied by rotating the specimen holder. At a moderate angle of incidence, only the specimen surface will be irradiated. However, by decreasing the angle of incidence and keeping the divergence slit width the same, both the specimen and the holder may be illuminated by the incoming X-rays. The scattering of X-rays from the specimen holder causes an increase in the intensity of background radiation. Conversely, on increasing the angle of the specimen to higher values, the irradiation area of the specimen decreases. A few diffractometers also employ variable divergence slits whose width can be adjusted to obtain a constant specimen irradiation area at all angles of incidence.

iv. *Specimen positioning in goniometer:*

The specimen holder should be positioned so that the specimen surface coincides with the goniometer axis. If the specimen surface lies above or below the goniometer axis, the measured value of Bragg's angle will be erroneous, and a shift in peak position is observed. The measured angle will be different from the actual diffracted angle. If the specimen is above the goniometer axis, it leads to a positive angular error, i.e., the peak positions will be shifted to higher 2θ values, and if the specimen is displaced vertically downward, it results in a negative angular error. Too much sample displacement from the goniometer axis may also cause asymmetric broadening of the peak at low 2θ values and result in poor resolution of the diffractometer. Careful specimen preparation and mounting on the specimen holder will minimise the effect of specimen displacement. Several refinement computer programmes, such as Rietveld and least-squares refinement programmes, have included a correction term for sample displacement errors.

v. *Takeoff angle:*

The angle between the plane of the target metal in the X-ray tube and the centre of the incident beam is known as the takeoff angle. The takeoff angle governs the effective width (W_e) of the X-ray beam, thereby influencing the peak and background intensities in the diffraction pattern. As the takeoff angle decreases, the effective width also reduces, as indicated in equation 2.15. A small takeoff angle produces a narrow X-ray beam and results in higher intensity but lower resolution. A high takeoff angle is also not preferred, as it reduces the resolution due to the increased number of out-of-focus X-rays striking the specimen. Generally, the takeoff angle is set in the range of 2°–8°.

$$W_e = W \sin \alpha \tag{2.15}$$

vi. *Scan range:*

The scan range is usually fixed based on the previous experience of the operator with the material to be examined. Since low-angle peaks are significant for phase identification, the scan range chosen should cover all possible low-angle peaks. For accurate qualitative phase analysis, it is recommended that the diffraction pattern contain around 50 peaks. However, if the material is highly symmetric, it may produce fewer peaks. The typical range of scanning angles using Cu $K\alpha$ radiation is from 5° to 65°. If the number of peaks is less than 50, then the scanning angle is increased.

vii. *Scan mode:*

Step scanning and continuous scanning are the two modes of scanning available in a diffractometer. The detector is moved at selected angular increments in step scanning until the last scanning angle is reached. The goniometer remains stationary at each angular position for a preset duration, during which the detector records the X-ray photons diffracted from the specimen. In continuous scanning, the detector is moved through increasing 2θ values at a constant angular velocity until the complete scan range is covered. Simultaneously, a counting rate meter counts the pulse rates, and a strip chart recorder displays the recorded diffraction data. Since the movement of the detector, counting of pulses and display of recorded data are all carried out simultaneously, the experiment time can be saved in continuous scan mode.

viii. *Step size:*

The difference between two consecutive diffraction angles 2θ preset during the experiment is known as the step size. In step scan mode, the step size and subsequent profile smoothing have to be carefully decided. The step size should be fixed such that sufficient data points (10–20) above full width at the half-height of the diffraction peak are obtained. A large step size and a high degree of profile smoothing suppress the peak intensity and result in low resolution. On the other hand, a small step size followed by a low degree of smoothing causes a shift in the peak position. Generally, a step size of 0.02° is used for crystalline materials.

2.11 EXPERIMENTAL PROCEDURE

The various procedures involved in the XRD study of materials are explained in the following sections.

2.11.1 Specimen Preparation

The sample to be examined has to undergo different treatment methods before being mounted on the specimen holder. Drying, grinding, sieving and dilution are a few steps in specimen preparation. Proper preparation is a prerequisite for obtaining accurate and precise peak positions and intensities in XRD patterns. For the qualitative analysis of a simple known material, a ground specimen without further preparation is placed in a specimen holder for examination. The orientation of particles is not given much importance in such cases. However, specimen preparation is more critical for quantitative analysis and analysis of complex materials, and it needs to be carefully executed. Various methods of preparation are available to minimise the preferred orientation effects of sample particles. Some of the methods are side and back loading of the specimen in the mould, using filler material, preparing a mixture of sample and viscous material, dispersing the sample in a binder, spray drying, etc. These methods are briefly described below.

i. *Grinding:*

Manual grinding using a mortar and pestle made of agate or ceramics and mechanical grinding using mills are the most common methods of grinding a sample. For efficient grinding in the mill, milling agents like acetone are used to obtain a sample of uniform particle size without overgrinding the softer particles. The grinding should not be continued for prolonged periods as it results in the agglomeration of particles and may affect the crystal structure of the material. The crystallinity of the material is also reduced when it is ground excessively.

The equipment used for grinding should always be kept clean to prevent contamination of the sample. Cleaning with a 50:50 mixture of nitric acid and water and then drying the mortar and pestle is necessary before grinding the sample in it. When the grinding of the sample is carried out in a mechanical mill, there is a possibility of the sample getting contaminated with the material of the balls used. Therefore, chemical analysis of the sample before and after grinding in a mill is performed to ensure that the sample is not contaminated.

ii. *Sieving:*

The ground samples are passed through sieves to remove large particles and break down particle agglomerates formed during grinding. Sieves of appropriate sizes in the range of 25–75 μm are used. The sieves should be clean to avoid contamination of the sample. Therefore, prior to screening the sample, the sieves are washed using alcohol and dried by passing gas such as nitrogen or helium at high pressure.

iii. *Spray drying:*

Spray drying is an effective technique to prevent the effects of the preferred orientation of particles in a sample. The particle size of the powder is reduced to under 10 μm by wet-grinding the sample in a vibratory mill. A slurry is prepared by mixing the ground sample with a quick-drying polymer solution consisting of a binder such as polyvinyl alcohol and a deflocculant. The slurry is then sprayed into a high-temperature gas zone, and during the flight, atomisation of the slurry occurs. The liquid part of the droplets will rapidly evaporate, and spherical agglomerates are formed. The spherical particles have the most random orientations and hence, remove the possibility of the particles being arranged in a preferred orientation during the mounting of the specimen in the holder. The major disadvantage of the spray-drying method is the expensive equipment and time required for preparation.

iv. *Use of filler materials:*

Filler materials are mixed with the powder sample to avoid the preferred orientation effect of the particles. Amorphous materials such as powdered glass, cork, gelatin and crystalline materials like corundum are used as fillers. The filler materials will surround the particles of the sample to ensure a random orientation of the particles. However, this method has the disadvantage that the sample may be contaminated by the filler material. The X-rays diffracted by the filler may appear as a background or as additional spectral lines in the diffraction pattern.

v. *Chemical extraction:*

Chemical extraction is the method of selectively dissolving certain phases in a multiphase sample using specific solvents. The method is adopted to avoid peak overlapping of phases that diffract X-rays at almost the same Bragg's angle. For example, a solution containing potassium hydroxide and sucrose selectively dissolves aluminate and ferrite phases. The residue after extraction contains silicates, carbonates and other minor phases. The steps involved in the process are:

- Prepare a solution of 30 g of KOH and 30 g of sucrose in 300 mL of water and heat the solution to 95°C.
- Add 9 g of cement to the solution. Stir the suspension for 1 minute, and keep the solution at rest for 30 minutes.

- Filter the solution and wash the residue in 50 mL of water and, after that, in 100 mL of methanol.
- Dry the residue at 60°C.

  Similarly, the use of a salicylic acid–methanol solution dissolves the alite, belite and lime present in cement. The aluminate and ferrite phases and several minor phases, such as MgO, calcium sulphate, and carbonates, remain in the residue.

vi. *Preparation of hydrated cement specimens:*

- A hydrated cement specimen has to be carefully prepared using the following procedure:
- The specimen (cement paste/mortar/concrete) is immersed in a lime solution for hydration.
- When the desired curing age is reached, the cement hydration is to be stopped. The hydration is arrested by removing the free water present in the capillary pores of the paste using direct drying or solvent exchange methods. It helps preserve the phase composition of the hydrated cement at that particular stage. The solvent exchange method involves treating the sample with organic solvents such as isopropanol, acetone, pentane, etc. The direct drying methods used are D-drying, oven drying, freeze-drying or vacuum drying. However, vacuum drying and oven drying are not preferred as they may cause dehydration of ettringite and AFm phases, which may result in an underestimation of the phase composition of such samples.
- The prepared sample is stored in a vacuum desiccator at low humidity. Before the examination, wet-grinding of the sample is carried out to obtain a powdered sample of appropriate particle size.
- Filter the ground cement paste slurry. Wash the residue with alcohol and, after that, with diethyl ether to remove any excess alcohol remaining in the sample.

  Cut slices of the hydrated cement specimen can also be used for examination using XRD. The surface of the slices is to be polished using sandpaper before placing them inside the diffractometer. Using cut slices instead of powdered samples has the advantage of reducing the preferred particle orientation effects. However, solvent exchange occurs faster on powdered samples compared to cut slices since the solvent exchange rate is governed by diffusion.

2.11.2 Specimen Mounting

Generally, a specimen may be loaded in a mould in three ways: bottom loading, side loading or top loading, depending on the type of the mould.

i. *Top loading:*

When the powder is loaded into a mould from the top, it is known as top loading. The powder should never be compacted on a smooth, flat surface since it causes preferred orientation effects. However, if necessary, gentle pressure against a rough surface can be applied to the powder surface while filling the mould. After loading, the top surface is made rough by cutting small groves using a sharp edge. The roughness of such a surface should be of the same order as the particle size of the sample.

ii. *Back loading and side loading:*

The following are the steps involved in the back-loading arrangement:

- Attach a frosted glass slide to the top surface of the mould using clamps or tape. The roughness of the surface of the slide facing the powder should be nearly the same as the particle size of the powder.
- Turn over the holder and carefully load the powder into the cavity. Only gentle pressure should be applied on the powder surface while loading. A rough specimen surface is necessary to avoid the preferred orientation effect of particles.
- Once the cavity is packed with powder, the excess powder on the surface should be removed in a single sweep using a blade or the edge of a glass slide. After that, a cover is placed over the powder surface.
- Turn over the mould so that the glass slide is on the upper side.
- Carefully remove the slide before the mould is placed inside the diffractometer and the specimen is exposed to X-rays.

In the side-loading arrangement, a special mould with an opening on the side is used. The procedure for loading is similar to the back-loading arrangement. However, it has the disadvantage that opening one side of the mould is not that easy compared to loading from the rear. Both back-loading and side-loading arrangements are better than the front-loading arrangement in minimising the preferred orientation of particles.

iii. *Zero background holder:*

Another sample mounting arrangement is to load the sample on a zero background holder or zero diffraction plate. The plate is prepared by cutting a crystal along a nondiffracting crystallographic direction and polishing it to extreme optical flatness. The main advantage of using this holder is that the X-rays striking it are not diffracted, hence the name zero diffraction plate. The diffraction pattern produced has no background noise and has a high signal-to-noise ratio.

iv. *Dispersion in a binder:*

In this method, the surface of a slide or plate is coated with a monolayer of viscous material such as vaseline, grease, machine oil, etc. A small quantity (a few milligrams) of the ground specimen is dusted directly on the thin layer of viscous material through a sieve. The excess powder remaining on the slide is removed by gently blowing air over the surface. Since the powder layer is very thin, the sample may be moderately transparent to X-rays and produce a diffraction pattern with unreliable peak intensities. Moreover, the scattering of X-rays by the viscous material may appear as an amorphous hump in the diffractogram. Due to these limitations, the method is recommended only when rapid mounting of powder with a minimum preferred orientation effect is required.

v. *Viscous suspension:*

If the specimen to be examined is a lightweight powder, a viscous suspension of the powder is prepared instead of directly loading the powder in the specimen holder. A chemically inert liquid with a low boiling point that does not dissolve the powder is used as the solvent. The prepared suspension is poured into the holder, and the excess quantity is removed by a single sweep using a blade. The solvent in the suspension is evaporated before the specimen is kept on the diffractometer for examination.

2.11.3 Stepwise Procedure for Operating XRD

A stepwise procedure for performing the XRD experiment is explained below:

Step 1: Follow the adequate specimen preparation methods explained in the previous section. If the specimen is moisture sensitive, the prepared specimen should be stored in a desiccator until it is time to perform the analysis.

Step 2: Take a sufficient quantity of the prepared specimen (powder) and fill the mould carefully. The even distribution of the powder is to be ensured by loosely overfilling the mould first and then gently tapping the specimen surface using a straight-edged tool such as a microscope slide or a spatula.

Step 3: Remove the excess powder from the mould by lightly scraping it with a straight-edged tool. The powder is lightly compacted using a rough-surfaced tool, such as a frosted glass slide. Use of a rough surface for compaction restricts the tendency of particles to arrange themselves in a preferred orientation.

Step 4: Check the alignment of the X-ray tube and the incident and receiving slits in the diffractometer.

Step 5: Press the unlock door button on the instrument and open the door. The operator should ensure that the busy light on the instrument is off before unlocking the door. While the door remains unlocked, the alarm light on the instrument will continuously blink.

Step 6: Place the specimen mould on the specimen stage of the diffractometer. Close the door of the instrument after correctly placing the specimen. The alarm light will stop blinking when the doors are locked.

Step 7: Carefully set the operating conditions, such as operating power, scan range, scan rate, scan mode and step size, using the instrument software. Slowly ramp up the voltage and current of the X-ray source in small increments to the desired output. It ensures that the filament in the X-ray tube has sufficient time to soak up the power and helps to extend the lifetime of the filament.

Step 8: Click the start button to run the scan once all the details are correctly entered. The specimen stage and the detector are automatically moved to the initial position set using the software. The shutter of the X-ray source will open, and the X-rays emitted from the source will strike the specimen. The busy light will be illuminated during the entire scan duration.

Step 9: During the scan, the X-ray source is stationary while the specimen stage and the detector are automatically rotated to different θ and 2θ positions, respectively (θ:2θ diffractometer arrangement). Diffractometers in which the specimen stage is stationary while the X-ray source and the detector are rotated at $-\theta$°/min and θ°/min, respectively, are also available.

Step 10: The scattered X-ray intensity as a function of the detector angle is recorded during the scan. The raw XRD pattern is simultaneously displayed on the computer monitor.

Step 11: Once the scan is complete, the shutter of the X-ray source will close automatically, and the busy light will turn off. Open the instrument door and remove the specimen from the specimen holder.

Step 12: The scan file is automatically stored in a folder on the computer. The raw diffraction data is processed to obtain the final reduced or digitised diffraction pattern.

2.12 DATA PROCESSING

Besides choosing appropriate operating conditions, the degree of processing adopted is important to extract accurate information from the experimental diffraction pattern. The various applications of XRD require the conversion of the raw diffraction data into reduced or digitised diffraction patterns. It is called a reduced pattern because data processing converts the large volume of raw data into concise digital data stored as a list of intensities and Bragg's angles. For this purpose, diffraction peak maxima are located, and the relative peak intensity is measured either as the peak height or the area under the peak. Figure 2.15 shows the steps involved in the processing of diffraction data. The smoothing of collected data, background reduction, and stripping of $K\alpha_2$ radiation are the initial steps involved in the data processing. The final steps include locating the peaks, calibrating 2θ, converting the peak data into d-spacing and storing the processed data for future use. The data smoothing and background removal processes should not be applied when the diffraction pattern is used for profile fitting or Rietveld refinement. This is because smoothing will distort the intensity-sensitive structural parameters of a crystal.

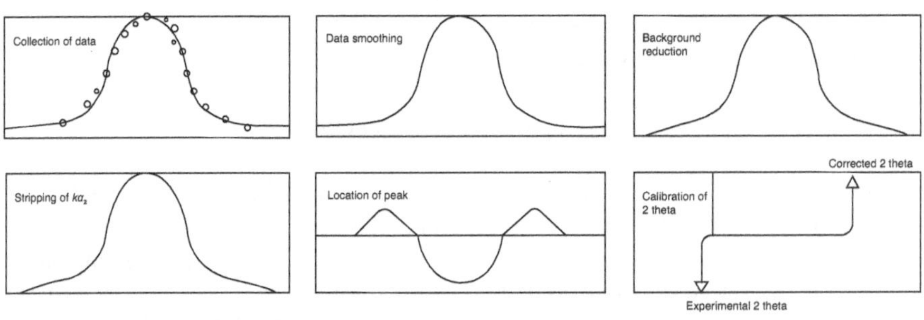

FIGURE 2.15 Steps involved in the processing of diffraction data.

2.12.1 Data Smoothing

Smoothing is a numerical conditioning process adopted to improve the profile shape and suppress the statistical noise present in the diffraction pattern, as illustrated in Figure 2.16. The presence of statistical noise may lead to the detection of false peaks. Firstly, to obtain a well-defined profile, an adequate number of data points must be considered by selecting a suitable step size. If a large step size is fixed for narrow peaks, the number of data points will be less. In such cases, smoothing improves the profile shape to a certain extent, but the peak intensity gets suppressed. On the contrary, smoothing broad peaks with a small step size reduces counting errors despite not improving the profile definition.

Box care smoothing is the most commonly adopted smoothing procedure. It involves taking the weighted average of intensities from a selected number of data points. The weight is highest for the point to be smoothed and decreases for neighbouring points. The degree of smoothing is fixed based on many factors, such as the number of data points, the sharpness, and the counting error associated with each peak. The operator may decide the number of data points considered for smoothing, or the software may automatically fix the number. If five data points are considered at a time, it is referred to as five-point smooth, and if it is seven data points, then it is called seven-point smooth. If the number of data points considered is too high, it may result in a loss of resolution and suppression of peaks.

2.12.2 Background Reduction

Processing a diffraction pattern with broad peaks superimposed by background noise is more complicated than one with sharp peaks and a low background. Hence, it is always important to keep the background as flat and low as possible. The background is either eliminated during preliminary data processing or added to the calculated intensity during profile fitting. In a typical pattern, the background is high at low Bragg's angles, minimum at the mid-angular range and gradually increases at higher Bragg's angles. The process of background reduction is carried out in two steps. Firstly, linearisation of the diffraction pattern is performed to remove the upward curvature of the pattern at low angles and the broad peaks due to diffused scattering. The second step is to select the points in the background either manually or

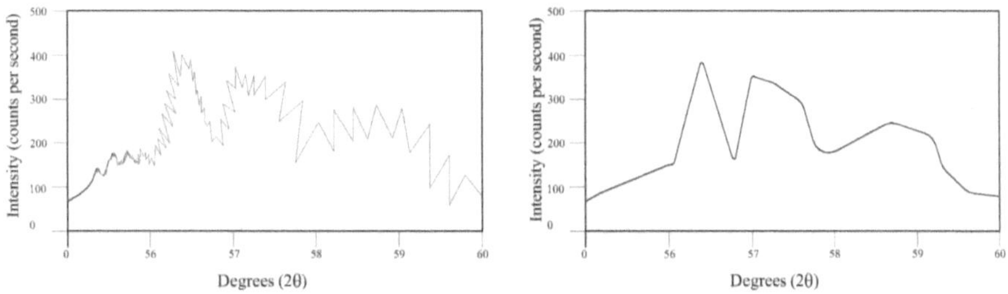

FIGURE 2.16 Smoothing process.

automatically for smoothing. The selection of the background threshold level has to be done such that a sufficient number of peaks are detected in the diffractogram.

2.12.3 $K\alpha_2$ Stripping

When the incident and diffracted beams are composed of $K\alpha_2$ radiation in addition to $K\alpha_1$ radiation, the diffraction pattern formed is a superposition of two sets of peaks due to the contribution from both radiations. The peaks of the $K\alpha_2$ component are located at slightly different 2θ angles compared to those of the $K\alpha_1$ component. At low Bragg's angles, the $K\alpha$ doublet is unresolved, whereas at high Bragg's angles, the doublet peaks are well separated. In the mid-2θ range, the overlapped peaks are partially resolved. The overlapping of peaks reduces the resolution of the diffraction pattern, and hence, $K\alpha_2$ elimination is necessary. However, the $K\alpha_2$ stripping method has a limitation at low 2θ values where the $K\alpha$ doublet is not resolved. The application of this method at low angles may reduce relative intensity by about one-third. The method may also lead to the reporting of fewer peaks in the pattern. Background reduction and $K\alpha_2$ stripping should be avoided if the processed diffraction pattern is used for profile fitting or Rietveld refinement. This is because the peak positions and intensities must be determined with good precision using these methods.

$K\alpha_2$ stripping is generally employed in automated data processing methods, whereas the weighted average of $K\alpha_1$ and $K\alpha_2$ radiations is used for manual methods. The elimination of $K\alpha_2$ peaks is based on the Rachinger method or deconvolution of the pattern using Fourier transforms. The Rachinger method assumes that the $K\alpha_1$ peak is identical to the $K\alpha_2$ peak and that the intensity ratios and wavelengths of the $K\alpha$ components are known. The diffraction peak intensity is proportional to the intensity of the two wavelengths in the characteristic spectrum. Hence, the peak intensity ratio is the same as the intensity ratio of $K\alpha_1$ and $K\alpha_2$ wavelengths, which is 2:1. Moreover, since the interplanar spacing of lattice planes diffracting the X-rays is the same for both radiations, the position of peaks can be related using Bragg's equation as given below. The contribution of $K\alpha_2$ wavelength in the diffraction pattern can be eliminated by applying these two conditions to all the points in the diffraction peaks as given by equation 2.16:

$$\sin\theta_1/\sin\theta_2 = \lambda_{K\alpha_1}/\lambda_{K\alpha 2} \tag{2.16}$$

2.12.4 Peak Search

The peak search can be performed in many ways, using manual or automatic peak search. Figure 2.17 shows the various methods of defining the peak position. The peak search is manually carried out either by locating the position of the peak maximum or by calculating the midpoint of the line joining the inflection points of a peak. The different phases in a material can be easily identified by obtaining the position of the peak maximum. Alternatively, the median angle of the peak is calculated such that the area under the peak to the left of the angle is equal to the area to the right. This method of manual peak search is known as the centroid method. The manual methods are more reliable but time-consuming. Nowadays, the most common data

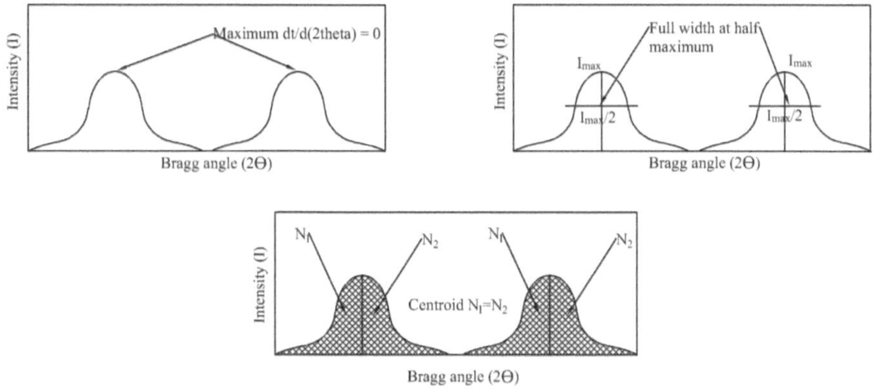

FIGURE 2.17 Methods of defining the peak position.

processing methods include smoothing, background reduction and $K\alpha_2$ stripping, followed by automatic peak search. The weak peaks and remaining background can be excluded during the peak search by setting a minimum intensity above which a detected peak will be considered. The profile scaling technique and the second derivative method are the two types of automatic peak search methods. In the profile scaling technique, peak shape parameters that match the best with actual data are determined using a strong, well-resolved peak. The obtained analytical peak shape is moved along the pattern, and the least-squares method is used to calculate the peak intensity.

In the second derivative method, after the initial data processing, the first and second derivatives of intensity with respect to 2θ are calculated to determine the peak position. The first derivative of the intensity function is zero at peak maximum, and hence, the peak can be located by computing the first derivative. However, the first derivative helps locate only resolved peaks in the pattern. To reveal the overlapped peaks in the pattern, the second derivative of the intensity function is calculated. The negative regions in the second derivative of the function represent the peak maxima. Since the sign of the second derivative changes at each inflection point, the full width at half the maximum of the peak can be obtained by measuring the range of negative region in the second derivative distribution.

2.12.5 Profile Fitting

When a diffraction pattern consists of closely overlapped peaks and asymmetric peaks, the most accurate method is profile fitting. It uses the non-linear least-squares technique to minimise the difference between the experimental diffraction pattern and the calculated pattern. The calculated pattern is the sum of the scaled profiles of the individual peaks detected and a suitable background function. Initially, approximate values of peak locations, peak shape parameters, and integrated peak intensities are determined. The peak locations are obtained by manual or automatic peak searches. The full width at half-maximum of the peaks is one of the peak shape parameters and is assigned a default value or estimated from a strong, well-resolved peak. These three parameters are varied during the least-squares fitting procedure to obtain the best fit.

2.12.6 Storage and Reporting

The processing converts the large volume of raw diffraction data into concise digital data stored as a list of peak intensities and Bragg's angles. The reduced diffraction data is stored on the computer and is easily accessible to all users. While reporting the XRD data collected, it is essential to mention the equipment model, diffractometer geometrical arrangement, X-ray wavelength, operating conditions, aperture of various slits used in the diffractometer, detector type, scan mode, sample type and preparation methods adopted, sample dimension, scanning parameters such as step size, scan range, measurement duration, etc.

2.13 ADVANTAGES OF XRD

The following are the advantages of using the XRD technique to characterise crystalline materials:

i. The sample size required for the experiment is very small.
ii. It is a non-destructive technique for material characterisation.
iii. It is a rapid and powerful technique for identifying unknown crystalline materials and phase compositions.
iv. The interpretation of recorded data is fairly straightforward.
v. Unlike conventional chemical analysis, the XRD patterns provide information on the state in which a substance exists. Conventional chemical analysis only detects the presence of various elements in a sample.

2.14 LIMITATIONS OF XRD

The following are the limitations of using the XRD technique:

i. The detector in a diffractometer records only the diffracted X-rays, which lie in the plane of the diffractometer circle. All the other scattered X-rays are not collected by the detector and remain undetected.
ii. For accurate analysis and identification of an unknown powder material, the sample should be homogeneous.
iii. XRD is less informative and accurate for materials with anisotropic particles and those with a non-uniform size distribution.

Hence, XRD has to be used in combination with other techniques, such as SEM, for precise analysis.

2.15 ISSUES RELATED TO XRD

Polychromatic nature, angular divergence of incident X-rays and spectral contamination of diffracted X-rays are the common issues that may affect the quality of the XRD data obtained.

2.15.1 Polychromatic Nature of X-rays

When a polychromatic beam of X-rays strikes a specimen, a set of lattice planes with unique interplanar spacing will scatter different wavelengths at different angles, satisfying Bragg's law. Hence, the observed diffraction pattern will be a superposition of patterns corresponding to the different wavelengths. The interpretation of diffraction patterns is more difficult in such cases compared to the interpretation of a pattern formed using a monochromatic X-ray beam of a single wavelength. Therefore, monochromatisation methods are employed to convert a polychromatic X-ray beam (with multiple characteristic wavelengths) into a monochromatic beam and also eliminate background radiation.

If a monochromator is placed in the path of the incident X-ray beam, it ensures that only the X-rays with the desired characteristic wavelength strike the specimen. The polychromatic radiation from an X-ray tube is converted into $K\alpha_1$ radiation, $K\beta$ radiation or $K\alpha$ doublet. The X-ray radiation generated from the X-ray source is filtered using a metallic foil or a monochromatic crystal. A crystal monochromator utilises a single crystal with highly reflective lattice planes. To obtain X-rays of the desired wavelength, the beam is directed at an angle θ with the plane, where θ is the Bragg's angle. A pair of crystal monochromators is suggested for perfect monochromatic radiation. The most common methods of monochromatisation are briefly explained below:

i. *β-filter:*

β-filters are used for monochromatisation of radiation produced in conventional X-ray sources only. The purpose of a β-filter is to reduce the intensity of the $K\beta$ spectral line and other unwanted radiations, which may affect the interpretation of the diffraction pattern. The β-filter allows the $K\alpha$ radiation to pass through it and selectively absorbs the $K\beta$ radiation. Suppose the filter is placed between the specimen and the detector; besides removing the $K\beta$ radiation in the diffracted beam, it helps to reduce the background due to fluorescent radiation from the specimen.

The variation of the absorption coefficient of materials as a function of X-ray wavelength was explained in Section 2.5.1. The position of the absorption edge of a material is the basis for choosing a material as a β-filter. Generally, a material whose K absorption edge lies between the $K\alpha$ and the $K\beta$ wavelengths of the target metal is chosen for the β-filter. The absorption line should lie above the $K\beta$ characteristic line and below the $K\alpha$ line, such that the $K\beta$ radiation is more strongly absorbed by the filter than the $K\alpha$ radiation. This condition is usually satisfied by choosing an element with one atomic number lower than the target metal. For instance, if copper

(Z = 29) is used as the target metal, nickel (Z = 28) is considered a suitable β-filter. Nickel will absorb 50% of Cu $K\alpha$ and 99% of Cu $K\beta$ radiation.

A few limitations of using a β-filter are that complete removal of $K\beta$ radiation is impossible and that the filtered beam has a considerable amount of white radiation. Moreover, a β-filter even reduces the intensity of $K\alpha$ spectral lines by a small factor. However, the intensity ratio of $K\alpha$ to $K\beta$ and $K\alpha$ to white radiation is significantly improved by using a β-filter. As the thickness of the filter increases, the intensity ratio of $K\alpha$ to $K\beta$ radiation in the transmitted beam becomes higher.

ii. *Crystal monochromators:*

A single crystal made of silicon, germanium, graphite, quartz, NaCl or LiF is commonly used as a crystal monochromator. A high-quality crystal is positioned in the incident X-ray path at a specific angle for the selective transmission of X-rays. The polychromatic beam leaving the X-ray source is divergent in nature and strikes the lattice planes of the crystal at various angles. However, the lattice planes oriented at a certain angle allow only X-rays of a discrete wavelength to pass through the crystal. The primary beam monochromators have the advantage of completely isolating the $K\alpha_1$ spectral line from the $K\alpha_2$ line. However, separating these two components is not usually done to avoid loss of beam intensity. Besides primary beam monochromators, the monochromatisation of the diffracted beam can be achieved by using diffracted beam monochromators. In this case, the crystal monochromator is placed after the Soller slits or the receiving slit in the diffracted beam path. The monochromator is oriented at a certain angle so that only $K\alpha$ radiation is diffracted by it. The detector is positioned to collect the desired wavelength of the diffracted beam. These monochromators are efficient in separating $K\alpha$ and $K\beta$ wavelengths and also in eliminating the fluorescent radiation emitted from the specimen material.

According to Bragg's law, each component wavelength of the incident beam is diffracted at a discrete angle. The wavelength of X-rays transmitted through the crystal depends on the lattice spacing and the incident angle. X-rays with shorter wavelengths, such as $K\beta$ radiation, are scattered at a lower diffraction angle, and those with longer wavelengths, such as $K\alpha$ radiation, are scattered at higher angles. The uneven spatial distribution of the diffracted wavelengths helps in the selective transmission of wavelengths by installing a slit in their path. The large separation of $K\alpha$ and $K\beta$ radiations helps in the removal of $K\beta$ lines. Crystal monochromators are also found to be efficient in eliminating most of the white radiation.

The quality of monochromatic radiation can be improved by arranging two or more crystal monochromators in a sequence. Parallel and angular configurations are the most common ways of arranging the monochromators. A flat monochromator is used in a parallel configuration, and it is known to eliminate the $K\beta$ radiation completely. On the other hand, in an angular configuration, the crystals are curved, and the orientation of the crystals leads to better separation of $K\alpha_1$ and $K\alpha_2$ radiation. However, this process is associated with high intensity losses, and hence, the $K\alpha$ doublet is accepted to maintain good intensity. Sometimes, one of the monochromators can be replaced with the specimen in these configurations.

iii. *Pulse height analyser and proportional detector:*

A β-filter allows a significant amount of white radiation along with strong $K\alpha$ and weak $K\beta$ radiation to pass through it. The size of the output pulses generated by the proportional detectors is proportional to the incident X-ray energy. It means that when X-rays of different wavelengths or energies strike the detector, output pulses of different heights are generated. A pulse height analyser is used to analyse and discriminate the pulse heights to sort out X-rays of different wavelengths entering a detector. Pulse height discrimination allows the detector to remove voltage pulses generated due to short and long wavelengths and select only X-rays of the desired wavelength for counting. Thus, β-filters are supplemented with a pulse height analyser to eliminate $K\beta$ and background radiation more efficiently. The resolution of the proportional detector influences the efficiency of white radiation removal.

iv. *Solid-state detector:*

Solid-state detectors such as Si(Li) detectors have good energy resolution and high energy proportionality. These detectors can be used to separate $K\alpha$ and $K\beta$ radiation. Hence, they are suitable for the monochromatisation of X-ray beams.

2.15.2 Angular Divergence

The X-rays produced from an X-ray source are diverging in nature. The angular divergence of the X-ray beam results in broad and asymmetric diffraction peaks. The irradiation width of the X-ray beam on the specimen surface depends on the divergence angle, incident angle, and goniometer radius. If the divergence angle of the emitted beam is large, then at lower incident angles, the width of the beam illuminating the specimen might be longer than the specimen length.

One or two divergence slits are placed between the X-ray source and the specimen to produce an in-plane collimated X-ray beam, as shown in Figure 2.18. The slit blocks the X-rays emitted at large divergence angles. Hence, only parallel or nearly parallel X-rays are allowed to propagate through the slit. The divergence angle of the collimated beam is given by equation 2.17:

$$\alpha_1 = (w + s)/d \tag{2.17}$$

where w is the width of the divergence slit, s is the width of focus of the source, and d is the distance between source and slit. Generally, by keeping s and d fixed, the width of the slit w is varied to adjust the divergence angle of the beam. The use of a narrow slit produces sharp diffraction peaks and better instrument resolution, implying that closely spaced peaks are easily distinguished. However, the narrower the slit width, the lower the intensity of collimated X-ray beam.

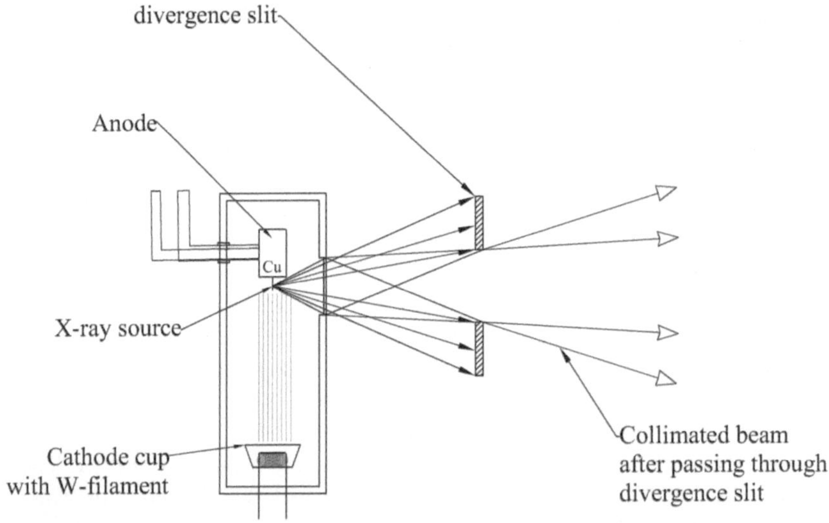

FIGURE 2.18 Provision of divergence slit to reduce angular divergence.

Additionally, a set of parallel and equally spaced thin metal plates called Soller slits are placed on the primary and diffracted beam paths to reduce the axial divergence of the X-ray beam (Figure 2.19). Axial divergence is defined as the divergence of the beam across the specimen along the goniometer axis. The angle of axial divergence can be varied by changing the length of the parallel plates l and the spacing between the plates d. Therefore, the in-plane divergence is reduced by using divergence slits, and the axial divergence is reduced by placing Soller slits in the X-ray beam path.

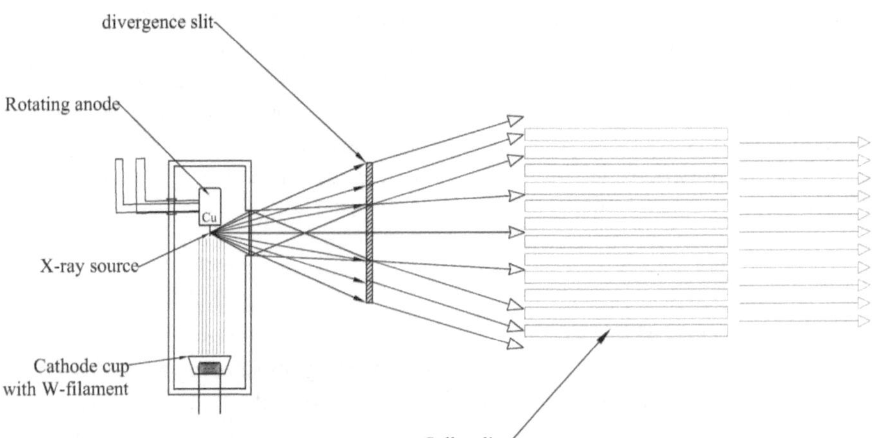

FIGURE 2.19 Collimation of X-ray beams using divergence slits and Soller slits.

2.15.3 Spectral Contamination

If a set of parallel lattice planes produce numerous diffraction peaks instead of a single diffraction peak, it is called spectral contamination. Even though monochromators are provided to remove unwanted wavelengths in the diffracted beam, a poorly aligned monochromator may lead to the appearance of additional weak peaks in the XRD pattern. $K\alpha_2$ radiation will be mostly present along with $K\alpha_1$ radiation in all the diffraction patterns, and both the radiations extensively overlap at low Bragg's angles.

The most common spectral contamination is caused by tungsten metal from the filament and copper from the anode in the X-ray tube. As the operational period of the X-ray tube increases, the contamination of the target metal by tungsten increases. This results in the appearance of weak peaks corresponding to tungsten in the diffraction pattern. The contamination check should be carried out regularly (every 6 months) by scanning a known specimen under normal operating conditions. The XRD pattern of the known specimen is then checked for additional peaks.

2.16 ANALYSIS AND APPLICATION OF XRD PATTERN

2.16.1 Diffraction Database

The widely used diffraction database is the Powder Diffraction File (PDF). It is a collection of standard XRD and crystal structure data for various substances such as elements, minerals, alloys, and organic and inorganic compounds. The interplanar spacing, Miller indices of lattice planes, density, optical information and corresponding relative intensities, which are valuable information for studying crystalline structures, are tabulated in the PDF. The wavelength of radiation, experimental conditions and specimen preparation are also mentioned in the database. The PDF is continuously updated by a nonprofit organisation named The Joint Committee on Powder Diffraction Standards (JCPDS), formed in 1969. The organisation was later renamed in 1978 as the International Centre for Diffraction Data (ICDD). Most modern X-ray diffractometers have PDFs stored on the computer and are easily accessible. The PDF-4 database supported by ICDD includes the crystal structure data from the Inorganic Crystal Structure Database (ICSD). The Crystallography Open Database (COD) and the American Mineralogist Crystal Structure Database (AMCSD) are the two alternative databases available for crystal structure data.

The data was printed in file cards in the early days, one card for each diffraction pattern. The cards are arranged into different groups based on the *d*-spacing. Later, the database became available on CD-ROM as well. Most modern diffractometers have these files stored on the computer, making the data easily accessible for the examiner. Over the years, as the number of substances added to the database increased, the number of cards increased, and directly searching the cards to identify a diffraction pattern became a cumbersome process. Hence, JCPDS introduced search manuals, and the cards were numbered chronologically or randomly rather than according to *d*-spacing values.

The ICDD has published two search manuals: the alphabetical index manual and the numerical index manual. In the alphabetical index manual, substances are arranged in alphabetical order of names. It is more beneficial for a quick search of diffraction data when the examiner has a rough idea of the phases present in the material. The numerical index manual is also known as the Hanawalt search manual. The Hanawalt search manual contains a list of substances and their eight most intense diffraction peaks. The diffraction patterns are sorted into different groups in decreasing order of the interplanar spacing of the most intense peak. The patterns are again sorted within each group in the decreasing order of the interplanar spacing of the second sharpest peak. Unknown materials can be identified using the search manual using the Hanawalt method. The database is also used for indexing diffraction peaks and calculating the lattice parameters of a known material.

2.16.2 Qualitative Phase Analysis

An XRD pattern is unique for a material, and hence, it can be used for qualitative phase analysis of materials. The experimentally recorded diffraction pattern of an unknown sample is compared with standard diffraction patterns in the PDF to identify compounds present in it. The search manual gives information on the position and intensity of eight strong peaks in the diffraction pattern corresponding to different *d*-spacings of a crystal. The first step in identifying an unknown material is to find the three most intense peaks in the experimentally recorded diffraction pattern and note down the corresponding interplanar spacing. Next, locate the group where the *d*-spacing corresponding to the strongest peak falls in the search manual. The *d*-spacing of the second-strongest peak determines the subgroup. The next step is to find the closest match for the *d*-spacing corresponding to the second and third strongest peaks. After obtaining a suitable match, the relative integrated intensities of the recorded XRD pattern are compared with the tabulated standard values in the PDF. If the values match well, the procedure is repeated for the next five most intense peaks in the diffraction pattern. An unknown sample can be identified as a particular material if all eight peak positions and intensities of the diffraction pattern match the eight peak positions and intensities specified in the PDF for that material. The matching procedure is more complex when the sample is a mixture of many phases. For mixtures, the strongest peak in the recorded pattern is identified, and different peak combinations are used to match the standard pattern. Once a match is obtained and a phase is identified, the peaks corresponding to that phase are eliminated from the recorded pattern. The process is repeated to identify the remaining phases.

A few points to be noted while carrying out the experiment for qualitative analysis are:

i. The particle size of the sample should be fine enough to give sharp and intense peaks in the diffraction pattern with a minimum background.
ii. Cu $K\alpha$ radiation is preferred to generate diffraction patterns since all the standard data available in the PDF was developed using Cu $K\alpha$ radiation.

One of the limitations of the paper search manual is its size, due to which computer-readable versions of PDFs were generated. The search and match algorithm in computers helps identify the major and minor phases present in a material. Among all the minor phases present, the result shows only a few of them on the screen. The analyst can choose which phases are to be displayed based on general experience or supplemental information about the material.

2.16.3 Quantitative Analysis of Phases

The peak intensity of a phase in a mixture depends on the concentration of that phase in the mixture and the absorption coefficient of the mixture. This is the basis of quantitative phase analysis. The peak height can be used as a measure of intensity if the peak height is proportional to the area under the peak. However, the intensity is distributed over a fairly wide 2θ range for broad peaks, and the peak height is not proportional to the peak area. In such cases, the integrated intensity is measured between two selected 2θ values instead of a direct measurement of peak height. This is because the peak broadening will affect the maximum intensity but not the integrated intensity. The third case is where the peak area is overlapped by peaks from other phases. In such cases, the profile fitting method using standard patterns or the internal standard method, in which the overlapped peaks are integrated together, is employed. The equation for the peak intensity of a phase in a mixture is shown in equation 2.18:

$$I_p = \frac{K_1 K_2 W_p}{\rho_p (\mu/\rho)_m} \tag{2.18}$$

where K_1 is a constant for an experimental setup, K_2 is a constant for each peak *(hkl)* of a phase in the mixture, W_p is the weight fraction of phase *p*, ρ_p is the density of phase, and $(\mu/\rho)_m$ is the mass absorption coefficient of the mixture. The mass absorption coefficient of a mixture is calculated as the sum of the products of the weight fractions and absorption coefficients of each phase in the material. In the above equation, the weight fraction of the phase and mass absorption coefficient of the mixture are the two unknowns to be determined in order to calculate the peak intensity.

2.16.3.1 Rietveld Analysis

In modern X-ray diffractometers, digital diffraction data of materials is stored on a computer and is easily accessible. The system can perform quantitative analysis of the entire diffraction data set rather than considering only a few intense peaks. The Rietveld method is one such method that uses the entire diffraction pattern for quantitative analysis. The method is more accurate and precise than the traditional methods involving the extraction of individual peak intensities. The measurement of an entire diffraction pattern in Rietveld analysis helps to overcome the issue of peak overlaps and the preferred orientation effect of particles. To obtain accurate results from the Rietveld analysis, a good diffraction pattern of a well-prepared specimen is required, and the phases in the pattern should be properly identified. The approximate crystal structure of all the phases of interest should be known to calculate the structural factors required in the Rietveld equation. The description of crystal structure for most of the phases is available in several crystal structure databases, such as ICSD, AMCSD and COD. An appropriate crystal structure model is selected from the database and then imported into a control file that serves as an initial input model for the Rietveld analysis.

The next step is to develop a theoretical pattern based on the selected crystal structure model and other initial input parameters. The profile shape parameters, unit cell parameters, scale factor, coefficient of background function, and specimen displacement correction are refined to match the theoretical pattern with the measured pattern. A least-squares fitting algorithm minimises the sum of weighted squared differences between the measured and calculated intensities in the pattern. The refinement is continued until the best fit between the theoretical pattern and the measured diffraction pattern is obtained. Finally, the quantitative information on phases is acquired using the scale factor for each phase in the material. The scale factor used in the Rietveld analysis is given as shown in equation 2.19:

$$S_p = K(W_p / \rho_p V_p^2 \mu_m) \tag{2.19}$$

where K is a constant for the experimental setup, W_p is the weight fraction of phase p, ρ_p is the density of phase, V_p is the unit cell volume and μ_m is the mass absorption coefficient of the mixture. However, K and μ_m cannot be determined from the experiment, due to which direct calculation of weight fractions using equation 2.19 is not possible. A solution to this problem is made by considering the sum of all the crystalline phases in the sample to be 100% under the assumption that the sample is composed of only those crystalline phases considered in the Rietveld analysis. The weight fraction of each phase is calculated as given in equation 2.20. A point to note is that calculating the weight fraction of phases in a sample containing amorphous phases gives only relative proportions of weight fractions.

$$W_p = \frac{S_p \rho_p V_p^2}{\sum_p^{n \text{ phases}} S_p \rho_p V_p^2} \tag{2.20}$$

2.16.3.2 Internal Standard Method

For the analysis of samples containing amorphous phases, the internal standard method is more recommended than Rietveld analysis. In this method, the XRD patterns of two specimens are determined and compared. The first is a sample of unknown phase composition, and the second is a mixture composed of a known quantity of an internal standard and the unknown sample. The internal standard is an arbitrarily selected material that is either purely crystalline or whose amorphous content is known. Preferably, the diffraction peaks of the internal standard should not overlap with the main peaks of the phases present in the sample. The most commonly accepted internal standards are corundum and zinc oxide.

The peak intensity of the X-rays diffracted from the phase p and from the standard s is written using equation 2.18. The two equations are divided to obtain the ratio of the peak intensities and eliminate the absorption factor from the equation. The internal standard method is based on the intensity ratio, as shown in equation 2.21.

$$\frac{I_p}{I_s} = k\,\frac{W_p}{W_s} \tag{2.21}$$

where I_p is the peak intensity of a phase in the first specimen, I_s is the peak intensity of the standard, W_s is the weight fraction of the standard in the second specimen and k is the slope of the internal standard calibration curve. Since the weight fraction of the standard is a constant, it can be said that the intensity ratio is a linear function of the weight fraction of phase p. The calibration constant is determined by mixing known concentrations of phase p with a fixed concentration of standard material and then calculating the intensity ratio. Finally, the concentration of each phase in the mixture is calculated. The absolute weight fraction of phases is determined using the expression equation 2.22:

$$W_p = W_s\,\frac{S_p\rho_p V_p^{\,2}}{S_s\rho_s V_s^{2}} \tag{2.22}$$

where S_s, ρ_s and V_s are the scale factors, the density and the unit cell volume of the internal standard, respectively. The obtained weight fractions are then compared with the weight fractions calculated using equation 2.20. If the weight fractions calculated by the internal standard method are higher than the latter, it indicates the presence of amorphous content or an unidentified crystalline phase. The apparent weight fractions are then corrected using the equation 2.23:

$$\frac{W_p}{W_{p,\,\mathrm{app}}} = \frac{W_s}{W_{s,\,\mathrm{app}}} \tag{2.23}$$

The total amorphous weight fraction is calculated as the difference between one and the sum of all crystalline weight fractions, as given by equation 2.24:

$$W_a = 1 - \sum W_p \tag{2.24}$$

However, if the value is lower, it indicates an error in the experimental procedure or data collection. If the two values are equal, it implies that the amorphous phases or unidentified crystalline phases are zero.

The internal standard method has the limitation that the addition of an internal standard may dilute the phases in the sample, and the accuracy of results may be lowered. Proper specimen preparation is required for the homogenous mixing of the internal standard and the sample. If the standard and the sample are ground together, care should be taken to avoid overgrinding the sample.

2.16.4 Determination of Crystal Structure

The crystal structure of an unknown substance can be determined by studying the XRD pattern of the sample. The size and shape of the unit cell are calculated by knowing the angular position of the peak in the diffraction pattern. Initially, the crystal system to which the structure belongs is assumed, and based on that assumption, pattern indexing is carried out. The indexing of the pattern is possible only when the crystal system is assumed correctly. Therefore, once the correct Miller indices are assigned to each peak, the shape of the unit cell can be determined from the crystal system. The number of atoms per unit cell is also calculated by knowing the density, the chemical composition, and the shape and size of the unit cell. Moreover, the position of atoms in a unit cell can be found by measuring the relative intensity of diffraction peaks.

2.16.5 Determination of Crystallite Size

Crystallites of smaller size do not have a sufficient number of parallel lattice planes to form a sharp diffraction peak. The resulting peak broadening effect can be used for the determination of mean crystallite size using the Scherrer equation. The width of the broadened peak at half the maximum intensity is measured in radians. The accurate determination of the size of crystallites is possible for diameters in the range of 100–200 nm. Since the crystallites of larger size have a sufficient number of planes to cancel out all the scattered X-rays except those scattered at Bragg's angle, the equation does not hold true for large-sized crystallites. Scherrer's equation, considering the peak broadening effect for small crystallite sizes, is shown in equation 2.25:

$$s = k\lambda / (B \cos\theta) \tag{2.25}$$

where s is the crystallite size, k is the Scherrer constant or shape factor (varies from 0.89 to 1.39), λ is the X-ray wavelength, θ is Bragg's angle, and B is the width of the broadened peak due to the particle size effect. B is calculated as the difference between the observed width of a diffraction peak at half the maximum intensity and instrumental broadening.

2.17 APPLICATION IN CONCRETE TECHNOLOGY

The XRD study is the most reliable technique for qualitative phase analysis of cementitious materials among all the characterisation techniques. The quantitative analysis of various phases using complete pattern fitting is also quite popular. However, due to the overlapping of diffraction peaks of several phases in a cementitious system, accurate measurement of peak intensity and determining the concentration of phases in the cementitious system are quite difficult. Hence, XRD is mainly used for qualitative analysis rather than quantitative analysis of phases. Cu $K\alpha$ radiation, a nickel β-filter combined with a monochromator system, and an aluminium crucible are commonly used for cementitious systems. However, copper radiation causes fluorescence of iron atoms present in the specimen, and the background intensity may submerge the low-intensity peaks, particularly in the low angular range. In such cases, cobalt may be used as the target metal in the X-ray tube, or other means to remove the fluorescent radiation may be employed. There are numerous applications of XRD in concrete technology, a few of which are described in the following sections.

2.17.1 Phase Identification of Anhydrous Cement

The various phases present in OPC cement clinker can be identified by qualitative analysis of XRD patterns. Special cements such as high-alumina cement, calcium sulphoaluminate cement, calcium fluoroaluminate cement and expansive cement can also be characterised using XRD data. The typical scan range for Cu $K\alpha$ radiation to examine a cement sample is from 7° to 70° 2θ and the standard step size used is 0.02° 2θ. The overlapping of diffraction peaks of phases is a major concern, and the various phases in the sample are easily distinguished by resolving all the overlapped peaks. The major peaks observed in an XRD pattern of cement clinker are monoclinic and triclinic forms of alite (C_3S), β-form of belite (C_2S), cubic and orthorhombic forms of calcium aluminate (C_3A) and orthorhombic form of calcium aluminoferrite (C_4AF). In addition, the pattern may show peaks corresponding to free lime, alkalis, sulphates, and periclase. The algorithm used for peak search shows only a list of suggestions for minor phases. The examiner must decide the minor phases to be detected based on prior experience with the material, supplemental information obtained by microscopic observations of the material or additional XRD studies on samples after selective dissolution treatment. The commercially used cement has gypsum added to the clinker produced in the kiln. Hence, the diffraction peaks of gypsum are detected in addition to the clinker phases. The lower angle region shows peaks corresponding to diffraction by C_4AF and gypsum phases in cement. The number of peaks detected for C_4AF depends on the cooling rate of the clinker. Rapid cooling of clinker results in fewer peaks and peak broadening due to the poor crystallinity of C_4AF.

If the cement sample is exposed to the atmosphere, portlandite is produced by the hydration of free lime in the cement. The portlandite formed later reacts with atmospheric CO_2 to produce calcite. Hence, it is imperative to ensure that the cement is not exposed to the atmosphere if the free lime content in cement is to be analysed using XRD.

2.17.2 Phase Identification of Hydrated Cement Paste

When cement reacts with water, several hydration products are formed. The hydration of silicate phases results in the formation of portlandite (CH) and calcium silicate hydrate (C–S–H), whereas ettringite and monosulphoaluminates are the main compounds formed during the hydration of C_3A and C_4AF in the presence of gypsum. Other hydration products formed are C_2AH_8, C_4AH_{13}, and C_3AH_6. Sharp peaks corresponding to all these phases except C–S–H are observed in the X-ray diffractogram of a hydrated cement paste. Since C–S–H is poorly crystalline, it does not produce a sharp peak; instead, a broad hump is observed.

2.17.3 Quantitative Phase Analysis of Cement

The quantitative knowledge of different phases in cement is significant in predicting the performance of cement concrete. The most common method of determining the cement composition is the Bogue calculation, aided by elemental analysis using X-ray fluorescence. However, the quantity of phases calculated using stoichiometric equations is different from the actual value. Another limitation is that the presence of minor phases is neglected in the Bogue calculation of phase composition. Estimation of phase composition is also possible by performing elemental analysis using energy-dispersive X-ray spectroscopy. However, this method is time-consuming as it requires the analysis of a large number of images to obtain a reasonably good estimate. Owing to the limitations of both methods, Rietveld analysis of XRD patterns is nowadays used for phase quantification. The Rietveld analysis involves the direct quantification of different phases in cement by measuring the peak intensities. It also has an advantage over other quantification methods, which involve measuring individual diffraction peak intensities instead of the entire pattern.

2.17.4 XRD Analysis of Concrete

Studying the phases in the interfacial transition zone of concrete and estimating the deterioration of concrete due to chemical attacks such as sulphate attack, thaumasite formation, etc. at the microstructural level is important for assessing the quality and long-term performance of concrete. For instance, crystal deformation, the nature of grain boundaries and grain size are a few factors that influence the alkali-silica reaction. The XRD study can be used to estimate the crystallinity index of silica, which helps determine the probability of an alkali-silica reaction in concrete. The assumption of Rietveld analysis is only valid if the amorphous content in the sample is negligible, as in the case of anhydrous cement. If the amorphous content is high, other methods that take into consideration the presence of amorphous silica should be used. This is the case for supplementary cementitious materials and hydrated cement. The internal standard method is the most common quantitative analysis method for samples with a high amorphous content. In addition, the crystal structure of phases present in concrete can be determined using diffraction data.

2.17.5 Monitor the Progress of Cement Hydration

The information from the XRD study can be used to monitor the hydration mechanism of hydrated cement. The evolution of different phases as cement hydration progresses can be studied using XRD data. The peaks corresponding to ettringite are detected at early curing days, and the peak intensity increases as the curing age of the paste progresses. The conversion of ettringite to monosulphoaluminate at later ages leads to the appearance of new peaks. The CH peaks are also detected within a few days of cement hydration. The appearance of the hydrated phases is accompanied by the disappearance of the clinker phases in the diffraction pattern. While analysing hydrated cement, the preferred orientation effects of plate-like crystals such as portlandite, monosulphoaluminates and ettringite with a needle-like shape should be minimised by appropriate specimen preparation or by applying corrections. While analysing hydrated cement paste or concrete at various ages, a direct comparison of the data obtained by Rietveld analysis will not give accurate results. As the curing age increases, the degree of hydration changes and the mass of solids also changes because part of the cement reacts with water to form C–S–H and other hydration products. Hence, normalisation of results is required for the comparison of data.

2.17.6 Assess the Reactivity of Pozzolanic Materials

XRD data can be used to assess the pozzolanic reactivity of pozzolanic materials such as fly ash, silica fume, slag, rice husk ash, sugarcane bagasse ash (SCBA), etc. The amount of amorphous silica present in pozzolanic material is a measure of the pozzolanic reactivity of the material. Hence, determining the amorphous silica content by quantitative analysis of the pozzolan is helpful to assess its reactivity. A broad peak or hump in the diffraction pattern of pozzolan indicates the presence of amorphous or reactive silica. Once the weight fraction of each major crystalline phase in the material is determined, the sum of the crystalline phases is calculated to estimate the amorphous content. However, this method is time-consuming, and the errors in the quantitative analysis of crystalline phases add up to the errors in the estimation of amorphous content. Figure 2.20 shows the X-ray diffractograms of hydrated OPC and OPC blended with 20% SCBA. A reduction in the peak intensity of CH can be clearly observed when 20% OPC is replaced with SCBA. The reduction in CH content indicates the consumption of CH in the pozzolanic reaction, and it can be related to the pozzolanic reactivity of SCBA.

2.17.7 Limitations of XRD Analysis of Cementitious Systems

i. Impingement of copper *Kα* radiation may cause fluorescence of Fe atoms in the system. Since the fluorescent radiation appears as a background in the XRD pattern, it may submerge the low-intensity peaks of C_4AF. Hence, Cobalt *Kα* radiation is preferred for the diffraction study of cementitious systems with high Fe content.
ii. Using the conventional method for quantitative analysis is a challenge due to the large degree of overlapping of peaks corresponding to major phases in cement. Hence, smaller peaks with a lower signal-to-noise ratio need to be considered for quantification. This leads to a significant reduction in the accuracy of the quantification.

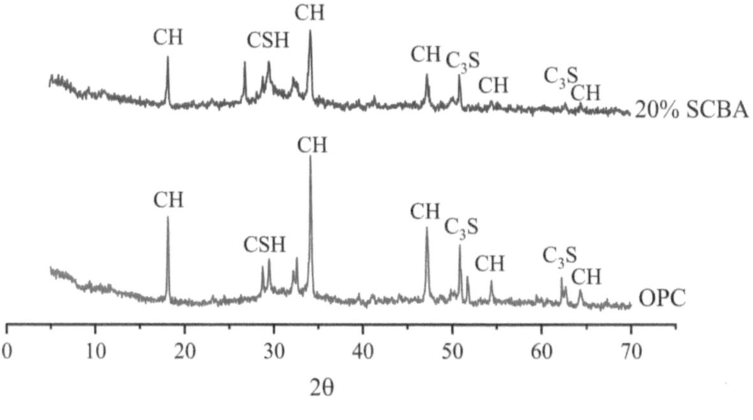

FIGURE 2.20 X-ray diffractograms of hydrated OPC and OPC blended with 20% SCBA.

iii. The XRD pattern of the hydrated cementitious system shows broad peaks of nanocrystalline phases such as C–S–H and amorphous phases of unreacted pozzolans. These phases are difficult to quantify using quantification methods such as Rietveld analysis. The overlapping of the broad peaks of C–S–H with peaks of crystalline phases such as belite may also cause a problem in the quantification of the crystalline phases.
iv. The stability of different polymorphic forms of major phases such as alite (C_3S) in cement depends on the solid solution of minor phases (alkalis, sulphate, etc.) in the crystal structure of the major phase. For instance, a solid solution of MgO in C_3S can stabilise the monoclinic form of C_3S. The variation in the crystal structure causes a change in the peak positions and relative peak intensities in the diffraction pattern. In such cases, the conventional quantitative analysis of measuring single peak heights may give erroneous results.
v. It is challenging to avoid the preferred orientation effect for certain cement phases such as calcite, gypsum, and ettringite. This is because they have a natural tendency to align in one direction due to the platelet-like or needle-like shape of the particles.

Therefore, it is highly recommended to compare the results of quantification analysis using XRD with the results of other analysis techniques such as scanning electron microscopy, thermal analysis, calorimetry or NMR spectroscopy.

2.18 APPLICATION IN OTHER CIVIL ENGINEERING FIELDS

The XRD technique has broad application prospects in the civil engineering domain.

2.18.1 Pavement Materials

Asphalt is a by-product of the petroleum refining process. The sticky, black and highly viscous liquid that remains after the distillation of petroleum is further processed to meet the desired specifications of an asphalt binder. The asphalt binder is mixed with crushed aggregates to produce asphalt concrete, which is used as pavement material on highways, runways and parking areas. Asphalt is composed of three major constituents: asphaltenes, resins and oils. The asphaltenes have a complex molecular structure consisting of large aromatic rings, long aliphatic chains, saturated paraffin chains and branched compounds, along with metals. XRD data can be used to characterise various materials used for pavement construction.

i. The X-ray diffractogram of asphalt provides information on the crystallite structure, composition, aromaticity and crystallite parameters such as interatomic distance and unit cell dimensions. The number of aromatic sheets, the number of aromatic rings in aromatic sheets, the diameter of aromatic sheets and the distance between aromatic sheets and aliphatic chains in asphalt can also be determined using the XRD technique. All these parameters are calculated by knowing the peak position, peak intensity and area under the peak in the diffraction pattern. Furthermore, the crystallite parameters can be used to determine other physical and chemical properties of asphaltene, such as molecular weight. A typical XRD pattern of asphaltene exhibits sharp peaks corresponding to gamma, graphene (002), (10), and (11) bands. The γ-peaks correspond to the aliphatic chains or condensed saturated rings and are observed at $2\theta = 17°–20°$. The graphene peaks representing the aromatic sheets are detected at $2\theta = 19°–25°$. The two other bands in the diffraction pattern appear at $2\theta = 43°–45°$ (10) and at $2\theta = 79°–81°$ (11). The (10) and (11) bands correspond to X-rays diffracted from the plane structures of aromatics. A broad peak at around 80° is also detected in the diffraction pattern.

ii. The ageing of asphalt binder is the main factor causing the deterioration of asphalt pavements. It directly affects the chemical composition of asphalt. The hydrogenation of aromatic rings occurs during ageing, and thereby, the aromaticity of the asphaltenes reduces. During the ageing of asphalt, other chemical reactions such as condensation, isomerisation, dissociation, dealkylation, etc. may occur. The XRD analysis can be used to study asphalt ageing.

iii. XRD analysis can be used to determine the crystal mineral composition of aggregates.

2.18.2 Soil

The soil mineralogy has a significant influence on the physical and chemical characteristics of the soil. The following are the various applications of XRD in geotechnical engineering:

i. The XRD data can be used to identify the various minerals present in the soil. The major mineral groups observed in the diffractogram of soil are amphiboles, carbonates, chlorites, feldspars, phyllosilicates, kaolins, micas, oxides, hydroxides, phosphates, pyroxenes, serpentines, sulphates, sulphides, and zeolites.

ii. Quantitative estimation of minerals present in the soil can also be carried out using XRD. The peak intensities in the diffraction pattern directly indicate the concentration of minerals in the soil sample. However, variation in crystal perfection, crystallite size, amorphous substances and chemical composition may affect the diffraction peak intensity. Hence, for accurate estimation, XRD analysis has to be used in conjunction with other methods such as TGA and elemental analysis.

iii. The XRD technique can be applied to obtain structural information such as unit cell dimensions, average crystallite size and lattice strain of the soil. The size of crystallites is important to understand the reactivity of minerals in soils. This is because as the crystallite size decreases, the surface area increases and the reactivity increases.

BIBLIOGRAPHY

Cullity, B. D., & Stock, S. R. (2001). *Elements of X-ray Diffraction*. 3rd ed. Pearson Education Limited, England.

Jenkins, R., & Snyder, R. L. (1996). *Introduction to X-ray Powder Diffractometry* (Vol. 138). John Wiley & Sons, New York.

Klug, H. P., & Alexander, L. E. (1974). *X-ray Diffraction Procedures: For Polycrystalline and Amorphous Materials*. 2nd ed. John Wiley & Sons, New York.

Pecharsky, V. K., & Zavalij, P. Y. (2008). *Fundamentals of Powder Diffraction and Structural Characterization of Material*. 2nd ed. Springer-Verlag, US. https://doi.org/10.1007/978-0-387-09579-0

Ramachandran, V. S., & Beaudoin, J. J. (2000). *Handbook of Analytical Techniques in Concrete Science and Technology: Principles, Techniques and Applications*. Elsevier, Amsterdam.

Scrivener, K., Snellings, R., & Lothenbach, B. (Eds.). (2016). *A Practical Guide to Microstructural Analysis of Cementitious Materials* (Vol. 540). CRC Press, Boca Raton, FL, USA.

Suryanarayana, C., & Norton, M. G. (1998). *X-ray Diffraction: A Practical Approach* (pp. 3–19). Springer, New York. https://doi.org/10.1007/978-1-4899-0148-4

Woolfson, M. M. (1997). *An Introduction to X-ray Crystallography*. 2nd ed. Cambridge University Press, New York. https://doi.org/10.1017/CBO9780511622557

MULTIPLE CHOICE QUESTIONS

1. Which of the following statements are true about Bremsstrahlung X-rays?
    i. They are produced due to the deceleration of the primary electrons.
    ii. They are made of a discrete characteristic wavelength.
    iii. They are produced due to the ionisation of an atom resulting from the ejection of an electron from the atomic shell.
    iv. They are made of X-rays of different wavelengths.
    a. Only i
    b. i and iv
    c. i and ii
    d. ii and iii
2. Which of the following statements are true about $K\alpha$ and $K\beta$ radiations?
    a. Electron transition from the L shell to the K shell causes $K\alpha$ radiation, and M shell to the K shell causes $K\beta$ radiation.
    b. $K\beta$ spectral lines are always stronger than the $K\alpha$ spectral lines.
    c. Electron transition from K shell to any outer shells produces K spectral series.
    d. The peak intensity of $K\alpha_2$ radiation is twice that of $K\alpha_1$ radiation.
3. Which of the following statements are true related to photoionisation?
    a. Photoionisation is associated with energy loss of incident high-energy particles.
    b. The energy of the incident particle should be greater than the excitation potential of the atom for photoionisation to occur.
    c. The release of characteristic X-rays accompanies the ejection of a photoelectron.
    d. All of the above.

4. How does the linear absorption coefficient vary as a function of X-ray wavelength for a specific element?
   a. Cubic parabolic variation
   b. Two linearly decreasing parts separated by a sharp discontinuity
   c. Two steadily decreasing curves (cubic parabola) separated by a sharp discontinuity
   d. Two steadily decreasing curves (parabola) separated by a sharp discontinuity
5. Which of the following phenomena may occur when a high-energy X-ray strikes a specimen?
   a. Elastic scattering
   b. Photoionisation
   c. True absorption
   d. All of the above
6. Which among the following falls under the category of incoherent scattering?
   a. Thomson scattering
   b. Compton scattering
   c. Photoionisation
   d. Both b and c
7. The phenomenon of re-emission of incident energy absorbed by a material is known as __________.
   a. Reflection
   b. Scattering
   c. Diffraction
   d. None of the above
8. Which among the following are the necessary conditions for X-ray diffraction to occur?
   i. Constructive interference of scattered X-rays.
   ii. Short-order periodicity of atoms in a crystal.
   iii. The angle of incidence of X-rays is equal to the Bragg's angle.
   iv. The path difference between scattered X-rays should be an integral multiple of the X-ray wavelength.
   a. i, ii and iv
   b. ii and iii
   c. i, iii and iv
   d. i, ii, iii and iv
9. For a θ:2θ system of diffractometer, which of the following statement is true?
   a. The position of the X-ray source is fixed, and the position of the specimen holder and the X-ray detector is varied during the scanning.
   b. The position of the X-ray detector is fixed, and the position of the specimen holder and the X-ray source is varied during the scanning.
   c. The specimen holder is fixed, and the X-ray source and the X-ray detector are rotated in opposite directions during the scanning.
   d. The position of the X-ray source and the X-ray detector is fixed, and the specimen holder is rotated about its axis to orient the specimen at different Bragg's angles.

10. Elements with an atomic number lower than chromium are not used as target metal in X-ray tubes because ________.
    a. The X-rays produced may get heavily absorbed by the tube window.
    b. The X-rays produced are in the shorter wavelength range.
    c. The emission will be dominated by white radiation.
    d. None of the above.
11. What is the function of Soller slits?
    a. Reduce angular divergence of X-rays
    b. Reduce axial divergence of X-rays
    c. Monochromatisation of X-rays
    d. Eliminate background radiation
12. A detector is said to be linear if _____?
    a. It is aligned at 180° with the specimen surface and the X-ray tube.
    b. The output pulses generated from the detector are linear.
    c. The rate of incidence of X-rays on the detector is directly proportional to the rate of output pulses generated.
    d. The amplitude of the generated voltage pulse is proportional to the incident X-ray energy.
13. Which one of the following is a proportional detector?
    a. Gas proportional detector
    b. Scintillation detector
    c. Solid-state detector
    d. All of the above
14. Scintillation is a phenomenon in which ______
    a. The X-rays absorbed by material are re-emitted as light photons.
    b. The high-energy electrons striking a material result in the emission of X-ray photons.
    c. The high-energy electrons striking a material produces X-ray fluorescence.
    d. The X-rays striking a material causes the emission of photoelectrons.
15. Which of the following statements is true about a semiconductor X-ray detector?
    i. The movement of electron-hole pairs generates electric pulses, which are proportional to the incident X-ray energy.
    ii. The detector must be kept at liquid nitrogen temperature always.
    iii. The detector must be maintained at vacuum pressure.
    iv. The Si(Li) detector has lower energy proportionality than a gas proportional detector.

    a. i, ii and iii
    b. i and ii
    c. i and iii
    d. i, ii and iv

16. The peak intensity in a diffraction pattern depends on ______.
    i. incident X-ray energy
    ii. depth of X-ray penetration
    iii. arrangement of atoms in a unit cell
    iv. size of the unit cell
    a. i and ii
    b. iii and iv
    c. i, ii and iii
    d. all of the above
17. **Statement A**: The diffraction pattern of amorphous materials shows a broad peak.
    Statement B: Amorphous materials have a periodic array of atoms with short-range order.
    a. Statement A is true, and Statement B is false
    b. Statement A is false, and Statement B is true
    c. Both the statements are correct, and B is the correct explanation for A
    d. Both the statements are correct, but B is not the correct explanation for A
18. Which of the following contributes to the background radiation in the diffracted X-ray beam?
    a. Diffused scattering of X-rays by the specimen
    b. X-ray scattering by the specimen holder
    c. Fluorescent radiation produced from the specimen
    d. All of the above
19. **Statement A**: Accurate diffraction pattern is obtained for specimens with platelet-like or needle-like shaped particles since they have a preferred direction of orientation.
    Statement B: Spherical shaped particles have the most random orientation and are ideal for obtaining accurate diffraction data.
    a. Statement A is true, and Statement B is false
    b. Statement A is false, and Statement B is true
    c. Both the statements are correct
    d. Both the statements are false
20. Which of the following information about a material can be obtained from an X-ray diffractogram?
    a. Crystal structure of the material
    b. The phase composition of the material
    c. Molecular bonds of compounds present in the material
    d. Both a and b
21. The spot size of an X-ray beam should not exceed the specimen length because ______.
    a. The X-rays falling outside the specimen area may be scattered by the specimen holder and produces background intensities in the diffraction pattern.
    b. The measured intensity of diffracted X-rays will be underestimated.
    c. Both a and b.
    d. None of these.

22. Which of the following is used as a filler material to avoid the preferred orientation effect of particles in samples?
    a. Gelatin
    b. Cork
    c. Corundum
    d. All of the above
23. Chemical extraction using salicylic acid–methanol solution results in selective dissolution of which of the following phases?
    i. Alite and belite
    ii. aluminate and ferrite
    iii. lime
    iv. carbonates and sulphates
    a. i and iii
    b. i, iii and iv
    c. ii and iii
    d. ii, iii and iv
24. Which of the following statements is true related to specimen preparation of hydrated cement?
    a. Vacuum drying may cause dehydration of ettringite and AFm phases.
    a. The preferred orientation effect is more pronounced for cut slices than powdered samples.
    b. The solvent exchange occurs faster on cut slices compared to powdered samples.
    c. All of the above.
25. Increasing the takeoff angle of an X-ray tube leads to ______.
    a. An increase in the effective width of the X-ray beam.
    b. A lower intensity of the X-ray beam.
    c. A decrease in the effective width of the X-ray beam.
    d. Both a and b.
26. The data smoothing of an X-ray diffraction pattern _______.
    a. Improves the profile shape.
    b. Increases the statistical noise present in the diffraction pattern.
    c. Separates $K\alpha_1$ and $K\alpha_2$ peaks in a diffraction pattern.
    d. Both a and b.
27. Which of the following are the manual peak search methods?
    i. Calculation of the midpoint of the line joining inflection points of a peak.
    ii. Locating the position of peak maximum.
    iii. Calculation of the median angle of the peak.
    iv. Second derivative method.
    a. i and ii
    b. i, ii and iii
    c. ii, iii and iv
    d. all of the above

28. **Statement A**: An element having one atomic number higher than the target metal is suitable for a β-filter.

    Statement B: An element whose K absorption edge lies between the $K\alpha$ and the $K\beta$ wavelengths of the target metal is chosen for the β-filter.
    a. Statement A is true, and Statement B is false
    b. Statement A is false, and Statement B is true
    c. Both the statements are correct, and B is the correct explanation for A
    d. Both the statements are correct, but B is not the correct explanation for A
29. Which of the following statements are true about monochromators?
    i. A crystal monochromator is always placed in the path of the diffracted X-ray beam.
    ii. A monochromator separates $K\alpha$ and $K\beta$ wavelengths and also eliminates background radiation.
    iii. Proportional detectors and pulse height analysers can be combined with a β-filter to eliminate $K\beta$ and background radiations.
    iv. A monochromator helps to reduce the angular divergence of an X-ray beam.
    a. i and iv
    b. ii, iii and iv
    c. ii and iii
    d. i, ii and iii
30. As THE width of the divergence slit reduces, _______.
    a. Sharper peaks with a better resolution are produced.
    b. The intensity of the X-ray beam increases.
    c. Axial divergence of the X-ray beam is reduced.
    d. Both a and b.
31. Which of the following methods is preferred for the quantification of amorphous phases?
    a. Rietveld method
    b. Hanawalt method
    c. Internal standard method
    d. Least-squares fitting method
32. Which of the following statements are true regarding XRD analysis of cement?
    a. Diffraction peaks of gypsum and C_4AF are mainly observed at higher 2θ values.
    b. A solution containing KOH and sucrose selectively dissolves alite and belite phases in cement.
    c. The major peaks observed in the XRD pattern of anhydrous cement are C_3S, C_2S, C_3A, and gypsum.
    d. All of the above.

33. **Statement A**: A broad hump extending over a 2θ range rather than a sharp peak is observed in the diffraction pattern of C–S–H.
    Statement B: C–S–H is a poorly crystalline material.
    a. Statement A is true, and Statement B is false.
    b. Statement A is false, and Statement B is true.
    c. Both the statements are correct, and B is the correct explanation for A.
    d. Both the statements are correct, but B is not the correct explanation for A.
34. Which of the following is not an application of XRD?
    a. Estimating damage of concrete due to chemical attacks.
    b. Assess the pozzolanic reactivity of pozzolanic materials.
    c. Detect cracks, fissures and voids in a concrete structure.
    d. Study the hydration mechanism and kinetics of cement.
35. Which of the following is a limitation of XRD analysis of hydrated cement?
    a. X-ray fluorescence of iron atoms in the sample appears as background when copper radiation is used.
    b. The accuracy of quantitative phase analysis is low due to the overlapping of major cement phases in the XRD pattern.
    c. The preferred orientation of calcite, gypsum and ettringite particles.
    d. All of the above.

1 b	2 a	3 d	4 c	5 d	6 d	7 b	8 c	9 a	10a
11b	12c	13d	14a	15b	16c	17c	18d	19b	20d
21c	22d	23a	24a	25d	26a	27b	28b	29c	30a
31c	32c	33c	34c	35d					

3 Scanning Electron Microscopy

3.1 OVERVIEW

A microscope is an equipment that reveals features of a specimen that are not visible to the naked eye. The resolution of a microscope and the contrast of various features in the image formed are the two factors that determine the quality of the image. Since the discovery of optical microscopes (OM) in the late 1500s, they have become quite popular in various fields. Electron microscopes (EM) were developed much later based on two significant findings: the wave nature of electrons and the use of a magnetic field to focus the electrons. EMs have higher resolution than OMs due to the shorter wavelength of the electron beam. The scanning electron microscope (SEM) and transmission electron microscope (TEM) are the two types of EMs.

In optical microscopy, the light beam falling on a specimen surface is reflected, transmitted, scattered, absorbed, or re-emitted at another wavelength. Similarly, in electron microscopy, the electron beam striking the specimen is scattered or absorbed, generating various characteristic signals. Secondary electrons (SEs), back-scattered electrons (BSEs), X-rays and auger electrons are the most important signals produced. These signals are captured by a detector and processed to extract useful information.

This chapter gives a brief description of the terminologies related to SEM and the instrumentation of an SEM. The various signals generated during the interaction of primary electrons with the specimen are discussed. The factors influencing electron imaging, including specimen characteristics and operating conditions, are also explained in detail. An in-depth explanation of the working principle of SEM and a stepwise procedure for conducting the experiment are given in the subsequent sections. A section focusing on the application of SEM in concrete technology and other civil engineering fields is presented. Issues related to imaging using SEM are also briefly described in this chapter.

3.2 COMPARISON BETWEEN ELECTRON AND OPTICAL MICROSCOPES

An EM and an OM have many similarities and differences. Both microscopes are used to obtain magnified images of objects that are not visible to the naked eye. A source of illumination, a condenser lens and an objective lens system are the major components of the two microscopes. Figure 3.1 shows the schematic diagrams of an OM, an SEM and a TEM.

DOI: 10.1201/9781032635392-3

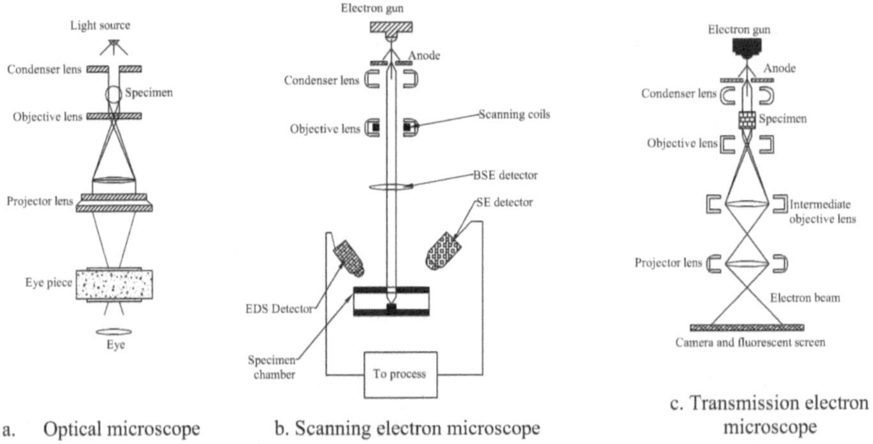

FIGURE 3.1 Different types of microscopes.

Some of the dissimilarities between the two microscopes are mentioned below:

i. Electron microscopy gives information about structural and elemental composition, whereas optical microscopy does not have that ability.
ii. EM should be operated at a high vacuum, while OM can be operated in any atmospheric condition.
iii. Energetic electrons produced by an electron gun are used as the source of illumination in EM, whereas visible light produced by a tungsten-halogen lamp is used in OM.
iv. The electron beam used in EM has a shorter wavelength (0.08 nm at a 20 kV accelerating voltage) than light (380–760 nm). A point to note is that a microscope's ability to resolve features in the specimen increases as the wavelength of the beam decreases. Hence, the image resolution and the ability to clearly distinguish small features are better for EM. OM has comparatively low-resolution power. The resolution of OM, SEM and TEM is 200, 1 and 0.1 nm, respectively.
v. Electrons move down the central axis in a spiralling action, whereas light rays travel down the optic axis as a parallel beam.
vi. An electron lens made up of copper coils encased in an iron casing is used for focusing the electron beam. The electromagnetic field generated in the lens is controlled to demagnify the electron beam. On the other hand, optical lenses made up of glasses are used in OM to ensure that the light beam moves down the optical axis as a collimated beam.
vii. Electron lenses are always convergent in nature, whereas optical lenses may be converging or diverging depending on the curvature of the lens surface.
viii. Focusing the electron beam on the specimen surface is done electrically by adjusting the intensity of the electromagnetic field in the electron lens. On the other hand, light rays are focused mechanically by moving the specimen stage vertically.

ix. EM can achieve a higher magnification of 90–800,000×, whereas OM can magnify images in the range of 10–2,000×. Hence, the microstructural characterisation of materials at different scales, ranging from micro to nanoscale, can be carried out using an EM.
x. EM has a greater depth of field and can be used for the detailed study of specimens with rough surfaces. The depth of field is small for OM.
xi. In EM, the electron-specimen interactions are mainly scattering, absorption (SEM) or diffraction (TEM). In OM, the image contrast is obtained by absorption or reflection of the light rays.
xii. The images formed by OM are directly visible to human eyes, whereas the signals formed after electron interactions with the specimen are processed to produce grayscale images.

3.3 SCANNING ELECTRON MICROSCOPY

The most widely used EM for examining the microstructure of materials is the SEM. In SEM, a high-energy electron beam is scanned across the specimen surface, and various signals generated from the illuminated area are captured by the detectors. The signals are processed to produce an SEM image or a micrograph with high resolution. The key feature of SEM is the three-dimensional appearance of the micrograph obtained in SE mode. Moreover, modern SEMs are often equipped with an energy-dispersive X-ray spectrometer (EDS or EDX) to determine the elemental composition of the specimen.

3.4 TERMINOLOGIES

Accelerating voltage	When the cathode (filament) in the electron gun is heated up, it emits electrons. The potential difference between the cathode and anode, which acts as the driving force for the electron beam, is called accelerating voltage (V).
Beam current	The electron charge per unit time incident on the specimen surface is defined as beam current or electron current (I).
Beam energy	Initial energy acquired by the electron beam as they enter the specimen chamber by the application of accelerating voltage is defined as beam energy (E). It is measured in electronvolts (eV).
Brightness	The current density (number of electrons emitted per unit area) per unit solid angle subtended by the electrons is called the brightness of the electron source. It gives a measure of the amount of beam current that can be focused on a given area of the specimen surface. For a constant current density, brightness is inversely related to the convergence angle of the electron beam.
Convergence angle	It is defined as half the angle of the cone of the electron beam converging into the focal point along the central axis (shown in Figure 3.2). The angle of convergence (θ) influences the spot size of the beam. The convergence angle can be controlled using lenses and apertures or by manipulating the working distance.

(Continued)

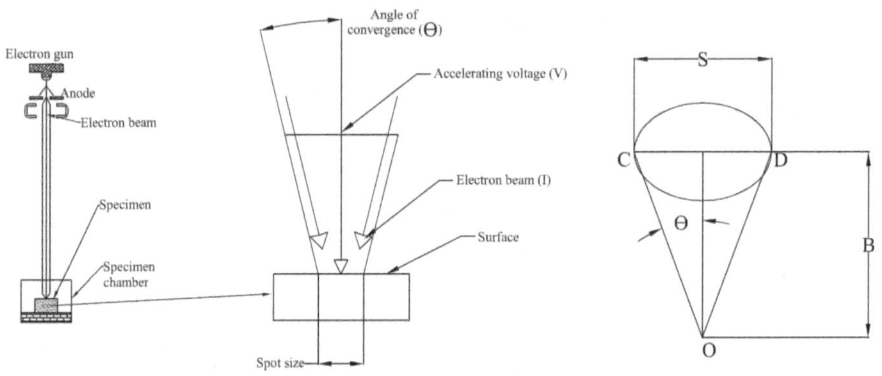

FIGURE 3.2 Schematic showing electron beam parameters.

Depth of field	The specimen length in the vertical direction along which all the specimen features are in good focus simultaneously is known as the depth of field. It indicates a microscope's ability to simultaneously maintain a sharp focus for a range of depths in the specimen. The regions of the specimen located outside the depth of field will be out of focus and appear blurry.
Edge effect	The phenomenon by which raised surfaces, ridges and edges (crest) appear brighter than broad flat regions (trough) in a SE image is known as the edge effect.
Elastic scattering	Scattering of the incident electrons from the specimen surface without any energy loss is known as elastic scattering. When the electrons are elastically scattered, their wavelength remains the same, but they are deflected from their initial path.
Focus	The electron beam leaving the final aperture of the objective lens converges at some point along the central axis of SEM, called the focal point or focus. The incident beam should converge at the specimen surface to obtain a sharp image with finer details. If the beam converges above or below the specimen surface, the beam diameter at the point of incidence will be larger, and the image will be out of focus. The focus of the microscope is adjusted by altering the current supplied and thereby controlling the magnetic field strength of the objective lens.
Image contrast	The contrast makes a feature of the specimen distinguishable (clearly visible) from other features and the background in an image. The difference in the intensity of signals generated at two locations gives rise to image contrast. The variation is because of the change in specimen composition, shape and orientation of grains at different specimen locations.
Inelastic scattering	The interaction of incident electrons with the specimen that results in a loss of energy is called inelastic scattering. The transfer of energy to the specimen when the electron beam hits it will generate signals that can be used to characterise the material.
Ionisation energy	The ionisation energy or electron-binding energy of an electron in an atomic shell can be defined as the minimum energy required to eject an electron out of an atom beyond the influence of the positively charged nucleus.

(*Continued*)

Magnification	The length of a pixel divided by the length of the corresponding picture element is termed magnification. In other words, it is the ratio of the scan length on the computer monitor to the scan length on the specimen from which signals are generated.
Microstructure	A structure at a micro-scale that is magnified and resolved using an EM for examination is known as the microstructure. The type, size, shape, quantity and distribution of phases in a specimen constitute its microstructure.
Picture element and pixel	The electron beam dwells on each discrete location of the specimen to generate a signal that is processed by a detector. Each of these locations is known as a picture element and the corresponding location on the computer monitor is called a pixel. The number of pixels on the monitor will be equal to the number of picture elements on the specimen.
Resolution	Resolution power is the ability of a microscope to distinguish two closely spaced points separately. The resolution limit is defined as the smallest distinguishable distance between two points. At a distance less than the minimum resolvable distance of a microscope, two points will appear as a single point in the image. The smaller the resolution limit, the higher the resolution of the microscope.
Specimen interaction volume	The volume inside the specimen where the incoming electron beam interacts with the atoms in the specimen is known as the specimen interaction volume. It indicates the maximum depth in the specimen through which the electrons penetrate before getting absorbed.
Spot size	It is the diameter of an electron beam when it impinges on the specimen surface, as shown in Figure 3.2. The final aperture of the objective lens determines the spot size (S), which in turn governs the resolution of the image.
Work function	The minimum amount of energy required for an electron to overcome the surface potential barrier and escape from the metal surface is defined as the work function of that metal.
Working distance	The distance between the pole piece (final aperture) of the objective lens and the specimen surface is known as the working distance (B).

3.5 INTERACTION OF ELECTRONS WITH SPECIMEN

The incoming electron beam interacts with the specimen at different depths within the specimen interaction volume to produce various signals, as presented in Figure 3.3. The specimen interaction volume depends on various factors, such as:

i. *Atomic number of the element:* If the atomic number is higher, most of the electrons will be absorbed, and the interaction volume will be smaller.
ii. *Accelerating voltage:* The higher the operating voltage, the greater the penetration depth of the electrons and the larger the interaction volume.
iii. *Angle of incidence of electron beam:* If the electron beam is incident at a greater angle, then the penetration depth of the beam will be reduced and the interaction volume will be smaller. An electron beam, which is incident vertically on the surface, creates maximum interaction volume.

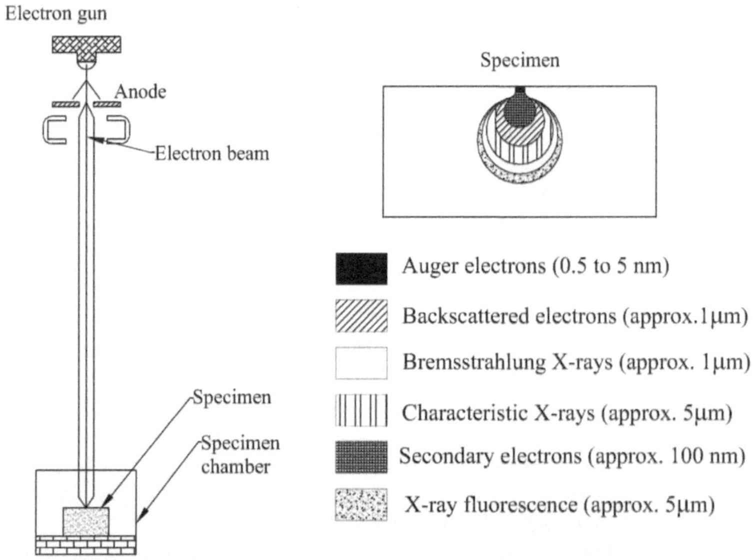

FIGURE 3.3 Various signals generated due to electron-specimen interaction.

The difference in the microstructure and orientation of different phases within the specimen leads to variation in the signals generated across the specimen. These signals are collected by electron detectors and processed to produce an image. The most significant signals emitted are secondary electrons (SE), BSEs and X-rays. These signals carry information about the specimen, such as morphological characteristics, chemical composition and crystalline structure. Some of the electron-specimen interactions may cause internal changes in the specimen instead of emitting electrons or X-rays. The generation of phonons and plasmons is an example of such specimen modifications.

3.5.1 Scattering

The incident electron beam deviates from its path upon striking the atoms in the specimen, and this phenomenon is called scattering. The scattering of electrons can be classified as elastic or inelastic scattering. The electron beam undergoes a series of elastic and inelastic scatterings within the specimen interaction volume until the energy of the incident electrons is completely dissipated. Sometimes the scattered electrons may re-emerge from the specimen as backscattered or transmitted electrons.

Elastic scattering occurs when the electrons are scattered without losing kinetic energy, whereas inelastic scattering is associated with energy loss. The collision of primary electrons (incident electrons) causes the electron beam to change its direction due to the repulsion from the positively charged nucleus in the atoms. The deviation of electrons from their initial path is higher in the case of elastic scattering. The inelastic scattering of the incident electrons is associated with transferring energy

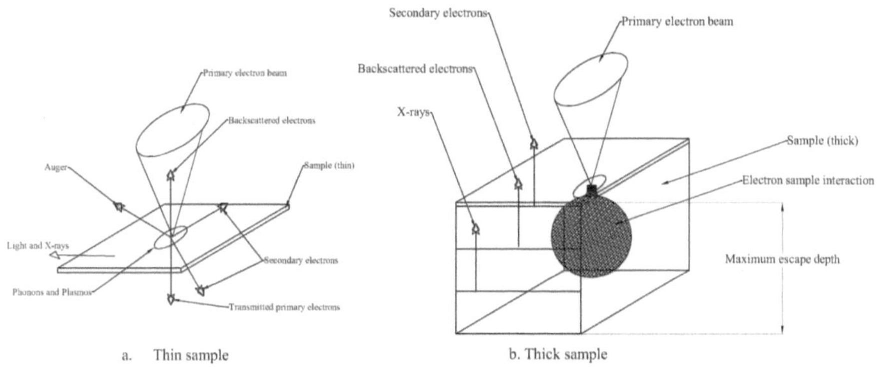

FIGURE 3.4 Signals generated for thin and thick specimens.

from the electron beam to the atoms. The energy transfer causes the excitation of loosely bound outer-shell (valence) electrons in the atoms. If the energy is high enough to overcome the electron-binding energy, the valence electrons are ejected from the atoms. These ejected electrons are called SEs. The inelastic collision may also release tightly bound inner shell electrons, leading to the emission of characteristic X-rays.

The thickness of the specimen affects the depth of penetration of electrons, the number of scattering events and the nature of signals generated, as shown in Figure 3.4. As the thickness of the specimen increases, the incident electrons will undergo multiple types of scattering, including elastic and inelastic scattering. With each inelastic scattering, the primary electron loses its energy and finally gets absorbed by the specimen upon complete dissipation of its energy. For denser specimens, electrons lose their energy faster, and the penetration depth of electrons into the specimen will be lower. If the specimens are less dense, then the scattering of electrons will be the predominant interaction instead of absorption.

Some of the important signals generated as a result of electron-specimen interaction are discussed below.

3.5.1.1 Auger Electrons

These are low-energy electrons emitted from the specimen at a shallow depth of 0.5–5 nm. Initially, the high-energy primary electron impinges on the atom to eject the inner shell electron (*K* shell), leaving a vacancy in the inner shell. The de-excitation energy released during the electron transition from a higher energy shell (*L* shell) to a lower energy shell (*K* shell) is sometimes absorbed by another electron in the *L* shell instead of being released as X-ray energy. A second electron is then ejected from the atom with characteristic kinetic energy, and the emitted electron is known as an auger electron. This results in two vacancies in the *L* shell of the atom, causing further electron transitions to fill the vacancies. The characteristic energy of auger electrons can be used to identify elements in the specimen, and this forms the basis of auger electron spectroscopy. An auger electron detector and an electron spectrometer are used to measure the number of auger electrons as a function of kinetic energy. The auger maps developed are very similar in appearance to the X-ray spectra of elements generated using X-ray spectroscopy.

3.5.1.2 Secondary Electrons

The SEs are generated by the inelastic collision of the incident electrons with the loosely bound valence electrons in the atom. When an electron impinges on an atom, the electron energy is transferred to the valence electrons, knocking them out of the valence shell of the atom. These electrons are called SEs. The SEs emitted from the specimen may be slow or fast, depending on their kinetic energy. The slow SEs are released from the valence shell with a very low energy of 5–50 eV. In contrast, fast SEs are ejected from the inner shells of the atom with up to 50% the energy of the primary electron (1–15 keV).

Only those electrons generated near the specimen's surface at a shallow depth of approximately 100 nm will have sufficient kinetic energy to overcome the potential barrier of the surface and exit the specimen. The low-energy SEs generated at greater depths in the specimen will not reach the specimen surface. This is because the energy of the SEs is attenuated as they travel through the specimen. In order to capture the low-energy SE leaving the specimen, a plate with positive potential is placed in front of the SE detector. The SEs emitted only from the top layers of the specimen are captured by the detector and contribute to the development of a SE image. Therefore, the SE signal produced can be used to obtain the topographical features of the specimen surface. The contrast and three-dimensional appearance of the SE image can be attributed to the variation in the number of SEs emitted from various features of the specimen. The contrast displayed in a SE image is known as topographical contrast.

Polishing and grinding the specimen surface is not required for SE microscopy. Hence, the surface will not be flat and have many crests and troughs. The SE detectors capture most of the SEs generated from the crest region, and these regions appear bright in the SE image. However, all the SEs generated from the trough regions may not reach the detector, and they appear as dark spots in the image. This is known as the edge effect. The angle of incidence of the electron beam also plays a role in determining the number and trajectories of SE signals generated. It varies with the local inclination of the specimen. The number of SEs is the lowest when the electrons are incident on the surface. At medium and high-tilt angles of the specimen relative to the electron beam, even the smallest surface irregularities can be observed clearly. It should also be ensured that the detector is placed facing the specimen surface to improve the efficiency of SE collection.

3.5.1.2.1 Types of the Secondary Electron Signals

SE_1: The SEs ejected from a shallow depth of the specimen when the incident beam impinges on the atom are classified as SE_1. The SE_1 signal is generated close to the region where the incident electron beam impinges on the specimen. The SE_1 signals are processed to produce a high-resolution image for studying the morphological characteristics of the specimen.

SE_2: The BSEs travelling through the specimen may collide with the SEs to undergo inelastic scattering and emerge from the specimen surface. The SEs formed in such a manner are known as SE_2. SE_2 is a low-resolution signal and possesses the

characteristics of the parent BSE signal. The strength of the SE_2 signal indicates the extent of backscattering in the specimen. The corresponding image represents compositional contrast rather than topographical contrast.

The production of SE_1 dominates for lower-atomic-weight elements due to the deflection of fewer BSEs from the specimen. On the contrary, an electron beam incident on a heavier atom produces many BSEs, and the possibility of generating a stronger SE_2 signal is higher. Hence, the contribution of the SE_2 signal to the total SE signal will be higher for heavier elements, and a region producing a stronger SE_2 signal will appear brighter in the image.

SE_3: These signals are indirectly generated when the BSEs strike the chamber walls or the objective lens pole piece of the instrument.

SE_4: These signals are generated by the inelastic collision of BSEs with the final aperture in the objective lens system.

All the SEs produced indirectly by the interaction of BSEs (SE_2, SE_3 and SE_4) are low-resolution signals and degrade the SE image quality.

3.5.1.3 Backscattered Electrons

BSEs are generated due to the elastic scattering of primary electrons striking a heavy mass, such as the nucleus of an atom. Owing to the high energy of BSEs, they can escape from greater depths (approximately 1 μm) of the specimen compared to SEs. They are scattered at angles ranging from 90° to 180° and are reflected off the specimen surface. A phase in a BSE image may appear bright depending on the number of electrons undergoing elastic collisions with the elements in that phase. In turn, the elastic scattering depends on the atomic weight and energy of the incident electron. The number of BSEs increases as the atomic weight of the atom increases. Moreover, the higher the atomic number, the greater the deviation of electrons from their initial path and the lower the penetration depth of incident electrons. The electrons will lose energy at a faster rate, resulting in a shallow specimen interaction volume with a hemispherical shape. On the other hand, the typical shape of the interaction volume when the electrons undergo elastic scattering after striking an atom with a low atomic weight is a teardrop.

BSE imaging is useful for the characterisation of materials. The contrast in the BSE images is created due to the difference in electron interaction with different specimen features. The amount and trajectory of reflected electrons result in image contrast. Another factor determining the contrast is the energy of the signal generated from the specimen. The resulting image conveys information about the structural composition of the specimen. The orientation of grains in the microstructure and boundary regions can be easily identified in a BSE image. BSE imaging is also helpful in identifying voids, cracks and fissures present in a specimen. The BSEs provide important information regarding material composition and crystallography. The image obtained after processing a BSE signal shows compositional contrast between different phases in the specimen. This variation arises from the difference in the surface topography and the change in reflectivity of different phases. Denser phases will have greater reflectivity due to a greater number of elastic collisions. For good compositional contrast, the specimen should be well-prepared and highly polished. A surface with a rough texture reduces the contrast and quality of the image.

3.5.1.4 X-rays

Inelastic scattering of primary electrons leads to the generation of X-rays. There are basically two types of X-rays. The Bremsstrahlung X-rays are produced from a depth of ~1 μm and characteristic X-rays from a depth of ~5 μm inside the specimen. The high-energy primary electrons are decelerated when they are deflected by a heavy mass, such as an atomic nucleus. The deceleration of primary electrons leads to radiation of energy in the form of X-rays. The generated X-ray continuum is known as Bremsstrahlung X-rays. In contrast, when an electron excited to a higher energy state returns to its lower energy state, X-rays with a series of discrete energies are radiated from the atom. Each element in a specimen emits characteristic X-rays corresponding to the difference in the energy levels of the atom. Characteristic X-rays are also produced when the vacancy created by the release of the secondary electron is filled by another electron from a higher energy level. Hence, the elemental composition of the specimen can be studied by processing the X-ray energy released.

3.5.1.5 X-ray Fluorescence

X-ray absorption is a phenomenon observed when X-rays emitted from an atom leave the specimen. As the X-rays pass through the specimen, if their energy exceeds the electron-binding energy of another atom, the energy can be transferred to the bound electron of the atom. This results in the ejection of the electron from the atom with a kinetic energy equal to the difference between the X-ray energy and the binding energy. The phenomenon is known as photoelectric absorption, and Figure 3.5 shows a schematic representation of the process. In photoelectric absorption, the primary X-rays give rise to low-energy, characteristic secondary X-rays, also known as fluorescent X-rays or X-ray fluorescence. Both primary characteristic and continuous X-rays can produce secondary fluorescent X-rays. The characteristic fluorescent X-rays are produced purely by primary characteristic X-rays, while the continuum fluorescence effect occurs due to primary continuum X-rays. However, since the X-ray continuum has low intensity over a wide range of energy, the fluorescent X-ray continuum will be limited.

3.5.1.6 Cathodoluminescence

When the electrons strike luminescent materials, the absorbed energy is re-emitted as light photons from the specimen surface. This phenomenon is known as

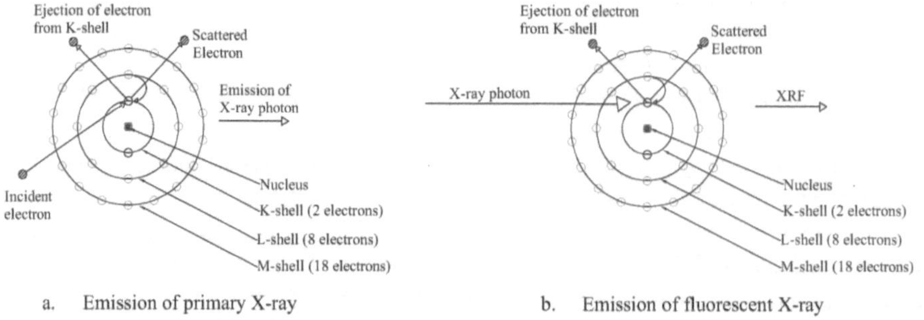

FIGURE 3.5 Schematic representation showing primary and fluorescent X-ray emission.

cathodoluminescence. A spectrophotometer or a photomultiplier sensitive to visible light is used for the detection of these signals. The colour of the photon emitted can be used to study the internal energy structure of the materials. It can provide information on impurities present in ceramics, identify trace elements in minerals, etc. The properties of semiconductor materials can be studied by bombarding the material with high-energy electrons, which leads to electron-hole pair formation and results in cathodoluminescence.

3.5.1.7 Absorbed Electrons

As the electron beam passes through the specimen, a part of its energy gets absorbed in the specimen. The net electron current remaining in the specimen after the emission of electrons is known as the specimen current. For semiconductor materials, the specimen current can provide information on the internal structure.

3.5.1.8 Transmitted Electrons

When thin specimens are used, some of the primary electrons may pass through the specimen and emerge from the specimen surface. Such electrons are called transmitted electrons and are used to obtain images of thin specimens in TEM. These electrons may or may not have interacted with the atoms in the specimen.

3.6 INFORMATION FROM SEM

The following information about the specimen under examination can be obtained through characterisation using SEM:

i. *Topography:* the arrangement of grains, defects, voids, cracks and other surface features can be observed using SE imaging. The shape of fractured specimens can be studied using SEM.
ii. *Morphology:* different phases present on the specimen surface can be identified using SE imaging. The shape and size of different phases of the material are visible in the micrograph. For example, different cement hydration products such as portlandite (calcium hydroxide), ettringite, and calcium silicate hydrate (C–S–H) can be distinguished well owing to the difference in their morphology.
iii. *Elemental composition:* identification and quantification of various elements that constitute the material can be carried out by using EDS.
iv. Compositional contrast and grain boundaries can be observed in a backscattered electron image.
v. *Crystallographic information:* arrangement of atoms in the material can be observed using electron backscattered diffraction.
vi. Pore size, shape and distribution can be studied using cryogenic techniques.

3.7 INSTRUMENTATION

The main components of an SEM are a column, a specimen chamber and a control system. Besides, several other components, such as a vacuum pump, a water chiller and various detectors, are part of the setup.

3.7.1 SEM Column

The SEM column is a long cylindrical body located above the specimen chamber. The presence of any gas inside the column causes energy loss from the electron beam. The gases may react with the electron source, causing electrons to ionise, producing random discharge and leading to instability of the beam. The transmission of the electron beam through the central axis may also be hindered by the presence of other particles, such as moisture. Hence, the column is always maintained under a very high vacuum. The column houses various components that generate and control the movement of the electron beam down the central axis of the column. Electron gun, condenser lenses, scan coils, stigmator coils and objective lens are schematically represented in Figure 3.6. The condenser lens converges the electrons generated from the electron gun to a spot below it; the converged beam may diverge again before being finally converged by the objective lens and focused onto the specimen surface.

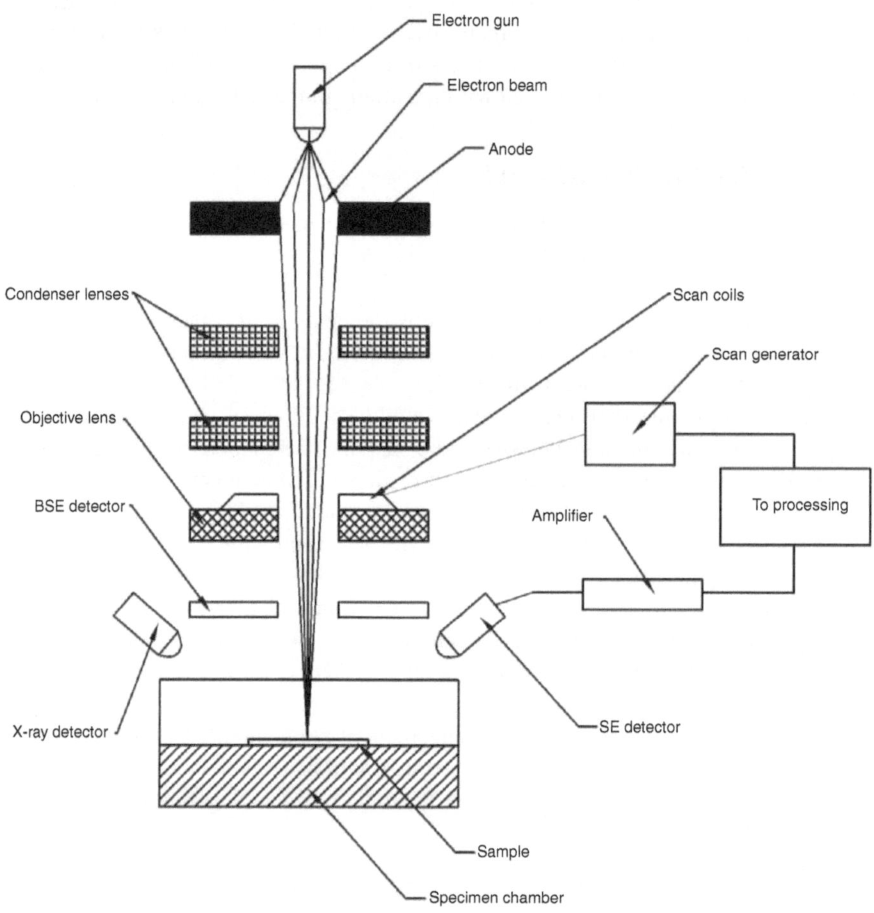

FIGURE 3.6 Schematic diagram showing various components of scanning electron microscope.

3.7.1.1 Electron Gun

The electron gun is an assembly of a cathode, which is the source of the electrons, and an anodic plate. It is located in the topmost portion of the SEM column, as shown in Figure 3.6. Figure 3.7 shows the parts of an electron gun in detail. The filament in the electron gun (which acts as the cathode) is connected to the positive pole, and the anodic plate is connected to the negative pole of a high-voltage source. When the electrons are emitted from the filament, they get scattered in all directions. The electrons travelling to the sides are repelled by the negatively charged Wehnelt cap surrounding the filament. The Wehnelt cap ensures that the electron beam stays in the central path and exits through the hole provided in the centre of the cap. The exiting electrons are collected in a space between the filament tip and the Wehnelt cap, which is called space charge.

The positively charged anodic plate attracts the electrons exiting through the Wehnelt cap towards it. The anodic plate is a metallic disc provided with a hole at the centre so that the electrons pass through it as a beam. The large potential difference between the cathode and the anode is the driving force that leads the electrons to accelerate down the column at a high speed. The potential difference, also known as the accelerating voltage, can be adjusted to control the acceleration of the electron beam. The typical operating range of accelerating voltage is from 1 to 30 keV. To achieve an accelerating voltage of 30 keV, the cathode is set at a negative potential of 30 keV, and the anode is set at zero potential. For concrete, 15–20 keV is commonly adopted.

Based on the mode of electron generation, thermionic and field emission (FE) are the two categories of electron guns. Thermionic filaments are resistively heated to a very high temperature for the generation of electrons. The filament tip offers high resistance to current flow and becomes the hottest part of the filament. The electrons are released when the tip becomes sufficiently hot, and the kinetic energy of free

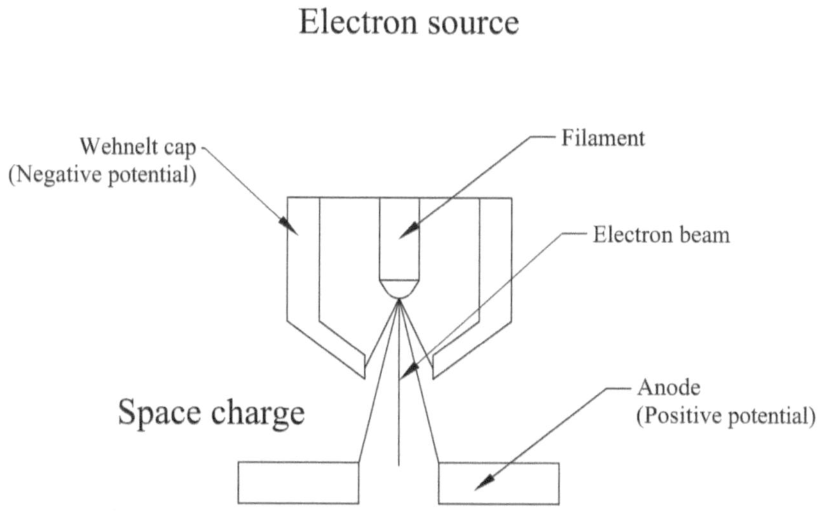

FIGURE 3.7 Electron gun.

electrons in the atom overcomes the potential barrier for the electron jump. Tungsten and lanthanum hexaboride (LaB_6) are the commonly used materials for thermionic-type filaments. They have low brightness and a short lifetime, but they have the advantage of not requiring a high-vacuum condition.

Alternatively, an FE-type electron gun is used as a source of electrons. FE gun generates an electron beam with 10,000 times higher brightness than a thermionic filament. The diameter of the electron beam generated by FE (5–10 nm) is much smaller than that produced by thermionic emission (10–20 μm) and it has the ability to concentrate a large amount of beam current in a small spot size. Thereby providing well-focused images with higher resolution. FE filaments have a long life since they are operated at low temperatures. Another advantage is the low-energy spread of an FE gun compared to a thermionic gun. However, an ultra-high vacuum in the range of 10^{-7} to 10^{-9} Pa is required for the FE gun to keep the filament tip clean from the electric field discharge. The basic principle of various electron guns is explained below:

i. *Tungsten filament:* a V-shaped thin tungsten wire mounted on an insulator is used as the cathode. Tungsten is an excellent filament material due to its high melting point, high tensile strength and low coefficient of thermal expansion. The filament is heated by passing an electric current, and when the heat is sufficient to overcome the work function of tungsten, electrons are emitted from the valence shell of tungsten atom.
ii. *LaB_6 filament:* LaB_6 filament is a crystal rod with a micro-flat, round or sharp tip from which the electrons are emitted. LaB_6 is an excellent filament material owing to its lower work function, higher melting point, higher thermal stability and longer lifetime compared to tungsten. However, LaB_6 has high electrical resistance, due to which it cannot be directly heated to emit electrons. The filament is mounted on a non-reactive graphite piece, which is resistively heated to raise the temperature of the filament. Since the filament is not directly heated to release the electrons, it is also an indirectly heated cathode. It is more expensive than tungsten filament and requires a high vacuum of 10^{-5} Pa.
iii. FE *gun:* A tungsten crystal with a sharp tip welded to a tungsten hairpin bent wire is used as the filament for a FE electron gun. Since FE can occur at low temperatures, it is also called cold FE. In an FE gun, a strong electrostatic field is applied at the tip of the filament to reduce the height of the potential barrier for electron emission. Consequently, the electrons can easily overcome the barrier without increasing their kinetic energy, and they are emitted from the filament tip.
iv. *Schottky* FE *gun:* It is a field-assisted thermionic electron gun; hence, is also known as a thermal-cathode FE gun. A tungsten crystal coated with a low-work-function adsorbate such as zirconium oxide is used as the cathode. The sharp tip of the filament is heated, and a high electrostatic field is applied to the heated metal to emit the electrons. Since the cathode is placed at a higher temperature under a higher vacuum, the electron beam current is more stable than that generated from cold FE. However, a cold-FE gun has superior performance owing to its longer lifetime, higher brightness and smaller energy spread.

3.7.1.2 Condenser Lens

The electrons ejected from the electron gun are passed through a pair of condenser lenses whose main function is to demagnify the electrons emerging from the electron gun into a thin beam. These are electromagnetic lenses, which consist of a copper coil enclosed in an iron casing. The magnetic field generated by passing current through the coil exerts a force on the electron beam. By increasing the applied current, the electrons can be focused closer to the lens, i.e., the focal length of the lens will be reduced. A strong focusing action causes the electrons to spread out after the convergence point, leading to the blockage of many electrons by the aperture before reaching the objective lens. This results in a smaller electron beam size and a lower beam current. On the other hand, decreasing the current applied to the condenser lens causes the electron beam to focus away from the condenser lens, resulting in greater number of electrons reaching the objective lens, thereby, increasing the spot size and beam current.

The magnetic field developed also helps to control the path of the beam and avoid electron deviation from the central axis. The direction of the magnetic field is fixed such that the force always tries to accelerate the electrons down the central axis. The magnetic field has two components: one in the radial direction, which is responsible for focusing the electrons, and the other in the axial direction, which causes the the electrons to spiral down the path. The electrons travelling through the axis are acted upon by equal field strength in all the radial directions, whereas off-axis electrons are subjected to unequal forces, which causes spiralling of the electrons about the axis. Moreover, the electrons away from the axis are subjected to greater force so that they are deflected more towards the centre, resulting in a focused beam. Since electrons are charged particles, the intensity of the magnetic field in the coils can be adjusted to vary the velocity of the electrons.

3.7.1.3 Scan Coils

The scan coils are electromagnetic double-deflection coils located in the objective lens assembly, as shown in Figure 3.8. These are controlled by a raster scan generator and are used to scan the electron beam over a specimen surface. The speed of scanning is controlled by varying the current applied to the coils. The scan rate influences the amount of noise in the image and the rate at which the image appears on the screen. Two sets of scan coils are used; the first set deflects the electron beam away from the central axis, and the second set deflects the beam towards the axis. Using these two scan coils, the electron beam is swept horizontally over the specimen surface from left to right (x-direction) and returned to the starting point. After the complete scanning of a row, it moves down a line (y-direction) and continues to scan from left to right (x-direction). In this manner, an entire rectangular area of interest is scanned, after which the beam returns to its initial position. The line-by-line scanning adopted by the scan coils is known as raster scanning. The specimen surface can also be scanned in a single line with the help of one set of scan coils, which is called line scanning. Each point on the surface is dwelt by the beam for a predetermined duration before moving to the next point in the line. If a large number of points are scanned, a realistic image of the specimen can be obtained.

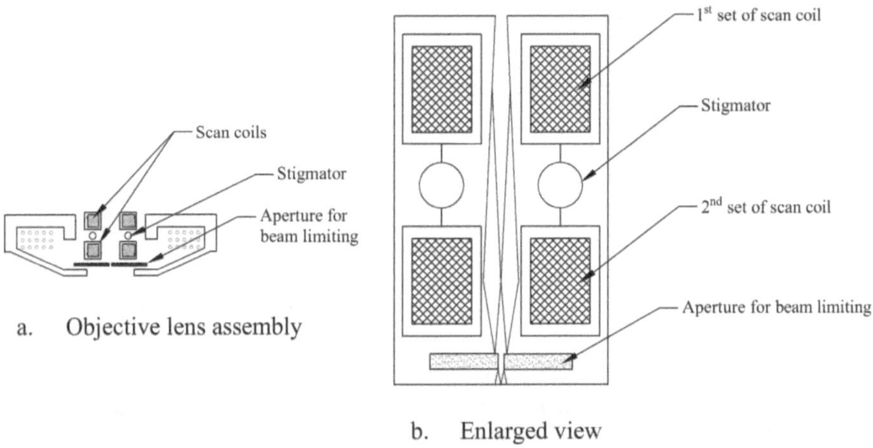

FIGURE 3.8 Objective lens assembly with scan coils, stigmators and aperture.

3.7.1.4 Stigmator Coils

A stigmator coil consists of an electromagnetic lens that is located near the pole piece of the objective lens. The stigmators are always aligned along the central axis of the SEM column. The coils are operated by adjusting the *x*- and *y*-stigmator knobs in the control panel. They are used to correct distortion of the electron beam shape in the *x*- or *y*-direction and thereby obtain a circular beam. If the electron beam is not circular but elliptical, the image appears blurred and stretched in one direction. The strength of the electromagnetic field in the coil is varied until a sharp image without stretching in any direction is obtained. Each time the stigmator is adjusted, the image needs to be refocused.

3.7.1.5 Objective Aperture

The aperture is a metal plate with a central hole (micron size) in the objective lens assembly (shown in Figure 3.8). It reduces the size of an electron beam by controlling the number of electrons and the convergence angle of the beam travelling down the column. The aperture blocks the stray electrons from passing through the objective lens and reaching the specimen surface. Hence, an aperture with a smaller radius aids in reducing issues related to lens aberration, resulting in better image resolution. Reducing the aperture size also reduces the electron beam current.

3.7.1.6 Objective Lens

The final focusing of the electron beam into the required spot size is accomplished by the objective lens. The size of the beam reduces by 100–5,000 times before striking the specimen surface to obtain useful images. The current passing through the coils in the objective lens is varied to control the magnetic field strength for adjusting the focal length of the lens. If the current is too weak or too strong, then the lens will be out of focus and the image obtained will be blurred.

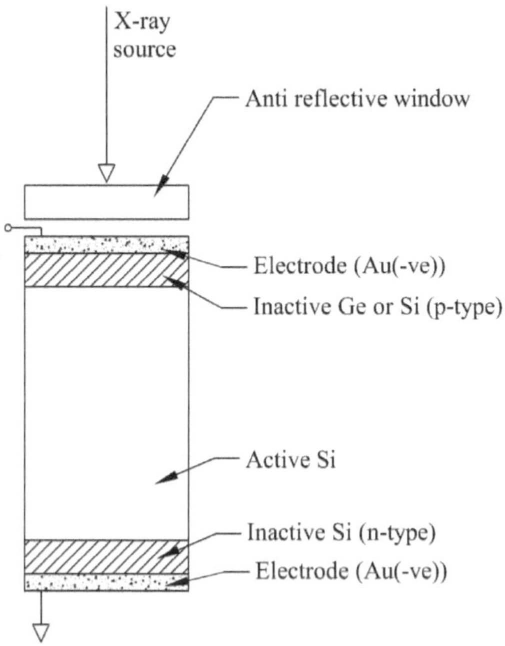

FIGURE 3.9 Energy-dispersive X-ray spectrometer detector.

3.7.2 Specimen Chamber

The specimen holding chamber, or specimen chamber, is located below the SEM column and is always kept separated from the column. The chamber houses many components of the microscope, including a specimen stage, an infrared camera and various detectors used for capturing the signals. A live image from inside the chamber is captured using the IR camera and displayed on the monitor. It helps the operator monitor the movement of the specimen stage and keep a check on the working distance to avoid collisions between the stage and the objective lens pole piece.

3.7.2.1 Specimen Stage

The specimen stage has a holder where the stubs containing the test specimen are loaded. Various forms of specimen holders are available, such as those with multiple positions for placing the stubs and cup-like holders for holding a single stub. The specimens are usually stuck to the surface of the stub using double-sided carbon tape, copper tape, etc. The specimen stage can be moved horizontally in the *x*- and *y*-directions as well as vertically (in the *z*-direction). Additionally, it can be rotated about the central axis by 360° and tilted at an angle of 15°–75°. The vertical movement of the stage is useful for adjusting the working distance and for lowering the stage to conveniently place the specimen on it. All these movements are performed either manually or are driven by a motor in modern instruments.

3.7.2.2 Detectors

Detectors are devices that are installed inside the chamber to collect the electrons or other signals leaving the specimen surface after the incident electrons impede the specimen. Different detectors are used to detect various signals such as SEs, backscatter electrons, and X-rays emitted from the specimen. The signals received are converted into an electric pulse by the detector. Finally, the micrograph (image) is displayed on the monitor after processing the received signal. The detectors are either permanently fixed inside the chamber or can be mounted and dismounted as per the user's requirements.

i. *Secondary electron detector:*

A SE detector, which is a scintillation detector, is used in SEM for detecting and processing SEs emitted from the specimen. The most common type of SE detector is the Everhart–Thornley (ET) detector, which is named after the designers who upgraded the previously available SE detectors. An ET detector is located inside the specimen chamber below the objective lens assembly. It is placed at an inclined angle with its front side facing the specimen to collect the maximum number of SEs. In the face of the detector, a positively charged collector grid, known as a Faraday cage, is provided. This is to capture the low-energy SEs (0–50 eV), including those that may scatter away from the direction of the detector.

Although an ET detector is used to detect SEs, it is also capable of collecting BSEs emitted from the specimen. The collector grid with a high voltage of +250 V has the potential to attract both SEs as well as low-energy BSEs in its line of sight. On the contrary, when the grid is connected to a voltage supply of –50 V, it will repel all the SEs and collect only line-of-sight BSEs. The ET detector has many advantages, including low cost, low noise, high efficiency in collecting even low-energy electrons, a large solid angle for collecting electrons, and a longer lifespan.

Mechanism of an ET detector

The electrons collected by the positively charged collector grid strike the scintillator placed inside the detector. A scintillator is made up of a luminescent material such as a silicon wafer, anthracene and zinc sulphide and is kept at a positive potential of 10–12 kV to attract electrons towards it. When the high-energy electrons bombard luminescent materials, they absorb the energy, and the electron in the atom gets excited and jumps to a higher energy state. On returning to the ground state, it emits energy in the form of photons with a wavelength in the visible spectrum range. This phenomenon is known as cathodoluminescence. The light is passed by total internal reflection through a plastic or glass pipe known as a light guide. The light guide directs the photons to a photomultiplier tube (PMT), which converts the light back into electrons. Initially, the electrons were converted to photons so that they could pass through a quartz window, which acts as a barrier between the evacuated specimen chamber and the PMT. The electrons emitted from the PMT are fed to an amplifier, which produces an amplified electric signal. The signal is passed to a video control unit and an image processing board before the image is displayed on the monitor screen.

The through-the-lens (TTL) detector is another type of SE detector that is placed above the objective lens. These detectors are used when the objective lens is strongly excited to produce high-resolution images, as in the case of FE SEM. They use the magnetic fields generated by the lens to capture the electrons. Since a shorter working distance is employed for FE SEM, the secondary electrons are efficiently trapped by the magnetic field. The BSEs travelling straight up through the axis are not attracted by the 10 kV potential of the scintillator in the TTL detector due to their high energy. Besides, the off-axis BSEs are not influenced by the magnetic field of the objective lens and are not collected by the TTL detector. Hence, the detector can effectively separate SEs from the BSEs. Another advantage is the exclusion of SEs generated when the BSEs strike the walls of the chamber and the pole piece of the objective lens. Since these SEs are produced far from the central axis, the magnetic field of the lens won't capture them efficiently.

ii. *Backscattered electron detector:*

A BSE detector is an annular-shaped detector installed directly above the specimen stage and is used for capturing the BSEs. Since most BSEs retain at least 50% of the energy of the incident electrons, passive detectors can be used for detection without any additional component for attracting the scattered electrons towards the detector. Typically, the detectors are placed either below the SEM column or at some distance inside the column. The maximum number of BSEs are collected by the detector when it is placed right above and normal to the specimen surface. If the BSE detector is placed at an inclination with respect to the specimen surface, for maximum collection efficiency, the specimen should be tilted towards the detector. The main two types of BSE detectors are scintillation-based and semiconductor-based detectors.

Mechanism of a scintillation-based BSE detector

A scintillation-based BSE detector works in a similar manner as a scintillation-based SE detector, except without a Faraday cage. In this case, the high-energy electrons directly strike the luminescent materials, causing the emission of photons. The photons are passed through the light guide to the photomultiplier, which converts the photons to electrons. The electrons are further processed to obtain a BSE image.

Mechanism of a semiconductor-based BSE detector

In semiconductor-based detectors, solid-state diodes are used to detect high-energy BSEs that penetrate the active region of the diode. The transfer of energy from the incident BSEs during inelastic scattering with the diode causes electrons to jump from the valence band to the conduction band of the atom. This creates a vacancy or hole in the valence band and leads to the formation of an electron-hole pair. The electrons in the conduction band are free to move, and by the application of an internal electric field, the electrons flow in one direction, resulting in current. This current is fed to an amplifier. The amplifier then generates an amplified signal, which is used to produce a BSE image.

The energy required for creating an electron-hole pair depends on the semiconductor material. For example, 3.6 eV of energy is required for electron excitation in a silicon atom. Hence, by knowing the ionisation energy per excitation of the semiconductor material and determining the number of electron-hole pairs, the energy of BSEs striking the detector can be measured. The higher the number and energy of BSEs striking the detector, the larger the number of electron-hole pairs created.

iii. *EDS detector:*

EDS detectors are used to detect characteristic X-rays that are emitted from the elements in the specimen. All the elements in the periodic table, ranging from beryllium to uranium, can be detected using EDS detectors. The first three elements in the periodic table, H, He and Li, cannot be detected due to the lack of electrons in the atomic shell to produce characteristic X-rays when hit by the incident electron. These detectors are installed very close to the specimen surface in such a way that most of the X-rays emitted are in the line of sight of the detector. A collimator tube placed in front of the detector acts as a limiting aperture and collects only those X-rays that originate from the specimen and restrains stray X-rays from entering. After the collimator tube, a pair of permanent magnets are placed to repel any incoming electrons. A very thin, opaque, anti-reflective window made of polymer coated with a thin layer of evaporated aluminium and supported with a silicon grid is kept after the permanent magnets. The evaporated aluminium coating is provided to protect the semiconductor crystal placed behind the window from visible light radiation. Additionally, the window keeps the inside of the detector separate from the specimen chamber and helps to maintain the vacuum condition inside the detector even when the specimen chamber is at atmospheric pressure.

Figure 3.9 shows the schematic diagram of an EDS detector. The X-ray photons striking the semiconductor diode made of silicon or germanium create electron-hole pairs through the photoelectric effect. The number of electron-hole pairs formed is proportional to the energy of incident X-ray photons. A bias voltage is applied between two gold contact plates (electrodes) attached to the opposite ends of the semiconductor. This causes the electrons and holes in the active region to move in opposite directions towards the collection electrodes.

The movement of the electrons creates current, and the flow of current between the electrodes is known as a charge pulse. The number of charge pulses produced is proportional to the energy of the X-ray photon hitting the diode. The higher the energy of X-ray photons, the larger the number of electron-hole pairs formed and the greater the current or charge pulse generated. Hence, by counting the number of charge pulses, the energy of X-ray photons emitted from the specimen can be obtained by multiplying the number of charge pulses with the energy required for forming one electron-hole pair. The signal obtained from the diode is fed to a field-effect transistor (FET). The FET amplifies the signal to increase its strength and reduce the signal-to-noise ratio. It converts the charge pulses into voltage steps. The size of a voltage step is directly related to the energy of X-ray photons striking the detector. The voltage step is finally transformed into a signal pulse using a pulse processor, and the X-ray spectrum is generated.

3.7.3 Image Display Unit

The output signal from the detectors is transferred to the display unit for producing a micrograph or an image. In recent years, liquid-crystal display has replaced cathode-ray tube as the display unit. The speed of refreshing the screen depends on the scan rate that is adopted. Even though a higher scan rate leads to a faster appearance of the image on the screen, a slower scan rate is preferred to obtain an image with good detailing.

3.7.4 Control System

The SEM is operated using a computer with software to set up a vacuum in the specimen chamber, operate the electron gun, control the specimen stage movements, store data, process signals, etc. Gun alignment control systems, condenser lens systems, scan coil systems, objective lens systems, astigmatism control systems and vacuum systems are some of the main computer control systems required for operating the SEM. The different knobs in the control console are used to adjust the various electron and imaging parameters, such as focus, magnification, brightness, contrast, accelerating voltage and beam current.

3.7.5 Vacuum System

The SEM column is kept at a high vacuum of around 10^{-4} Pa, whereas the vacuum in the specimen chamber is slightly lower and is in the order of 10^{-3} Pa. The vacuum pressure created in the column is even higher (10^{-6} to10^{-7} Pa) for FE SEMs. It is necessary to maintain a high vacuum inside the instrument for several reasons. Firstly, the tungsten filament used for emitting electrons may oxidise and burn out at very high temperatures in the presence of air. Another reason for the high-vacuum requirement is to provide a clean, dust-free and moisture-free environment for the operation. The electrons may otherwise interact with the dust particles in the column and chamber, causing electron scattering before reaching the specimen. Hence, a high vacuum ensures that the electrons do not stray from their path and increases the mean free path of the electrons. The moisture and other contaminants, if present inside the chamber, may deposit on the specimen surface. If not pumped out, they will significantly affect the imaging of fine surface details.

A vacuum system consists of vacuum pumps, valves, airlocks, vacuum gauges, etc. Different types of vacuum pumps are used to maintain adequate vacuum in the instrument. Combining two or more pumps increases the speed and efficiency with which the column and chamber are evacuated. The initial pumping is carried out using a rotary mechanical pump, also known as a rough pump. This pump removes the air from the chamber through a pipe connected to an outlet. An initial vacuum of 10^{-1} Pa is obtained using a rough pump. The final vacuum (10^{-3} to 10^{-4} Pa) inside the chamber is achieved by employing a turbo-molecular pump or an oil diffusion pump. The turbo-molecular pumps are preferred over the diffusion pumps since the latter may contaminate the chamber with oil vapour due to probable back pressure. The very high vacuum required for FE SEM is achieved using two ion-getter pumps. A Pirani gauge, or a cold cathode gauge, is the most commonly used pressure gauge to measure the pressure level in the chamber.

For obtaining good-quality images, the chamber should be kept in a low vacuum condition of 10^{-1} to 10^{-4} Pa pressure. However, it is difficult to create high-vacuum conditions for a wet specimen due to the evaporation of moisture from the specimen during the vacuum pumping of the chamber. This may result in the clogging of the exits in the microscope. Hence, for studying wet specimens, a variable-pressure SEM or Environmental Scanning Electron Microscope (ESEM), is employed.

3.8 WORKING PRINCIPLE

The primary electrons are emitted from the electron gun when the filament is heated or excited using an electric field. The accelerating voltage is varied to control the speed and intensity of the electron beam. The electrons are accelerated down the central axis and focused as a thin beam on the specimen using several electromagnetic lenses. The electron beam released from the gun passes through a pair of condenser lenses, which focus the electrons into a narrow beam. The objective lens aperture placed between the condenser lens and the objective lens further reduces the beam size by regulating the convergence angle and the number of electrons passing through it. The final adjustment in spot size and focusing of the beam on the specimen surface is done by controlling the magnetic field strength of the objective lens. The scan coils connected with a raster scan generator are used for scanning the electron beam across the specimen surface in a raster-like pattern. During scanning, the centre of the beam is focused on a specific location on the specimen surface for a fixed duration. Most SEMs have a feature called frame or line integration, where the image of the same area or line is captured multiple times and averaged to obtain the final image with less noise. The depth of penetration of the beam into the specimen depends on the accelerating voltage and density of the specimen and typically lies in the range of 100 nm–5 μm.

The interaction between the incident electrons and the specimen results in the generation of different signals such as SEs, backscatter electrons and characteristic X-rays. These signals, produced at discrete locations on the specimen, are captured by the corresponding detectors. The detector produces an amplified electric signal corresponding to each location (x, y) and is stored in a data matrix of three dimensions (x, y, I_j). I_j is the signal information from various sources, such as SE, BSE and X-ray, collected during the scanning of the specimen surface. The output of the detector is then processed using an image processing board. Finally, the output (micrograph or spectrum) of the scanned area is displayed on the monitor screen. The micrograph can be divided into a square grid of picture elements or pixels and each pixel corresponds to the signal generated from a distinct point on the specimen surface. Depending on the number of electrons collected from each location, it appears bright or dark in the image.

3.9 INFLUENCING FACTORS

The major factors, including specimen parameters and instrument operating conditions, influencing the characterisation of materials using SEM are explained below.

3.9.1 Specimen Characteristics

i. *Atomic weight:*

The atomic weight of the specimen material governs the image resolution. For the same beam current, materials with lower-atomic-weight elements will have a larger specimen interaction volume than materials with heavier elements. If the interaction volume is projected to the surface and the projected area is larger than the picture element at a particular magnification, the sharpness (resolution) of the image reduces. The image appears blurred because the signals generated by the adjacent picture elements overlap.

ii. *Specimen size:*

Specimens of 3–20 cm diameter can be kept inside the specimen chamber. It should be ensured that stage motion cannot lead to a collision with the lowest part of the instrument (pole piece of the objective lens). Important conditions for specimen mounting are secure fixing (particularly for high-tilt position), good electronic contact with the specimen holder and stage (to inhibit charge accumulation), and vertical positioning to ensure that the optimum focal range of the electron beam is reached.

iii. *Specimen topography:*

The smoothness of the specimen surface governs the angle of incidence of the electron beam on the surface and, thereby, influences the number of SEs emitted from the specimen. A higher percentage of the primary electrons are incident on the inclined portions of the surface or crest region such as ridges, edges and slopes and thereby emit a greater number of SEs. The SE detectors capture most of the SEs generated from the crest region, and these portions appear bright on the micrograph. On the other hand, the surface features in the trough regions may appear shadowed or dark due to the absence of a clear path for the SEs to reach the detector. The appearance of crest regions as bright and trough regions as dark is known as the edge effect. The surface orientation has an effect on the appearance of dark or bright colours on rough surfaces. Hence, the specimen should be tilted and rotated at various angles to capture the surface topography appropriately.

3.9.2 Operating Conditions

i. *Accelerating voltage:*

The typical operating range of accelerating voltage is 1–30 kV. Generally, a high accelerating voltage is preferred as it increases the beam current and reduces the electron beam size, thereby increasing the image resolution. It also leads to greater penetration of the electron beam into the specimen.

However, the accelerating voltage has to be fixed based on the nature of the specimen under examination. For denser specimens such as steel, an electron beam with a higher accelerating voltage helps capture the material characteristics better and results in higher image resolution. On the other hand, striking a low-density specimen with a high-energy electron beam results in a large interaction volume. This increases the contribution of low-resolution signals such as SE_3 and BSE to the SE signal, which leads to a lower signal-to-noise ratio. Moreover, a higher accelerating voltage causes a higher edge effect, surface charging and damage to low-conductive materials. Hence, for softer and non-conductive materials such as biological systems, low accelerating voltages are preferred for high-resolution imaging and to avoid damage to the specimen. A low accelerating voltage reduces the specimen interaction volume, and SE_1 and SE_2 signals are produced near the specimen surface. The SE_2 signal also contributes to high-resolution imaging.

ii. *Beam current:*

A high beam current results in a good signal-to-noise ratio and smooth images but compromises image resolution. It can also cause damage to soft materials. A low beam current is preferred to obtain good image resolution. However, a very low beam current produces a coarse image that fails to reveal the details of surface features. For capturing images at low magnification, very high resolution is not required; hence, high beam current and large spot sizes can be used to generate signals. Therefore, an optimum beam current is carefully chosen based on magnification, specimen type, etc.

iii. *Convergence angle:*

The convergence angle of the electron beam is controlled by varying the electromagnetic field of the lens system and aperture diameter. The lenses demagnify the electrons into a fine beam, and the aperture prevents the off-axis electrons from reaching the specimen. The convergence angle can also be decreased by increasing the working distance. A smaller convergence angle results in the focusing of the beam into a smaller spot size and decreases the beam current. The small spot size thereby leads to better resolution of the image.

iv. *Depth of field:*

The depth of field is governed by the convergence angle of the electron beam, which depends on the diameter of the objective lens aperture and working distance. A smaller aperture and a longer working distance result in a smaller convergence angle and a larger depth of field, as depicted in Figure 3.10.

v. *Magnification:*

Higher or lower magnification is obtained by decreasing or increasing the scan length on the specimen as the display length on the monitor is fixed. Magnification is varied by adjusting the magnetic field strength associated with the scan coils. Higher magnification is required to observe the features of the specimen more closely. However, imaging at higher magnification

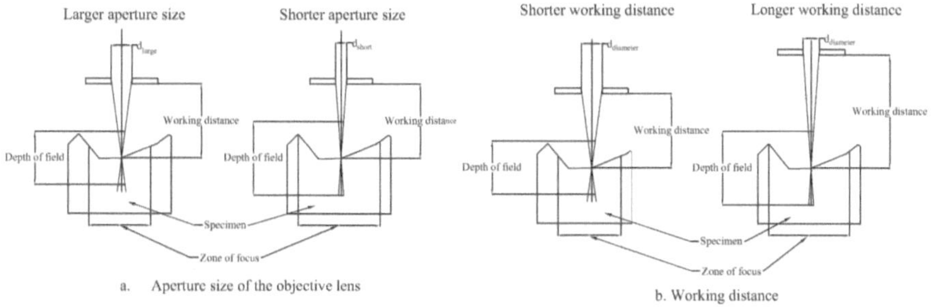

FIGURE 3.10 Effect of aperture size and working distance on the depth of field.

reduces the sharpness and resolution of the image. This is because the overlapping of signals will be more extensive due to a smaller picture element at a higher magnification.

vi. *Scan rate:*

The scan rate influences the signal-to-noise ratio and the rate of appearance of the image on the screen. A slower scan rate results in a higher signal-to-noise ratio and an image with good quality, but the refresh rate on the screen will be slow. However, image drift can happen during slow scanning, which results in the blurring of the images. Fast scan rates result in a lower signal-to-noise ratio because the number of primary electrons interacting with each picture element is less, resulting in a weaker signal. Moreover, very fast rates of scanning are not advisable since certain features on the specimen surface might not get captured.

vii. *Spot size:*

The resolution of an SE image is governed by the strength of the SE signal received by the detector, which in turn depends on the spot size. The spot size is mainly governed by the electron beam size at the source, the degree of convergence by lenses, and the spherical aberration of the lenses. For a given beam energy, a smaller spot size produces an SE signal with a lower signal-to-noise ratio due to a lower beam current. Hence, it results in a grainy image with high resolution. Larger spot sizes and higher beam current produce sharper images due to a higher signal-to-noise ratio. However, the overlapping of signals from adjacent regions is a problem for beams with large spot sizes, degrading the image resolution.

viii. *Tilt of specimen surface:*

The specimen surface is at times tilted at a particular angle to observe specimen features otherwise not prominent or to obtain stereomicrographs (3D reconstruction of surface topography). The specimen interaction volume is at its maximum when the electron beam falls perpendicular to the specimen surface. As the degree of tilt or angle of inclination of the specimen with respect to the beam is made lower than 90°, there is a chance of a greater number of electrons escaping the specimen surface and thereby reducing the specimen interaction volume. Another issue due to the tilting of the specimen surface is the variation in spot size as the electron beam scans

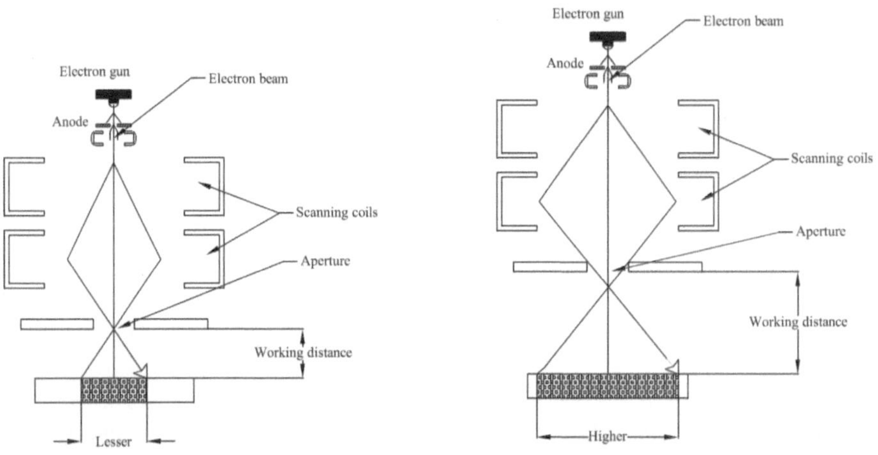

FIGURE 3.11 Effect of working distance on spot size.

from one end to the other end of the specimen surface. Dynamic focusing is an effective way to ensure that the spot size remains constant along the vertical axis. Dynamic focusing is achieved by adjusting the focal length of the objective lens as the electron beam scans over the tilted surface.

ix. *Working distance:*

A shorter working distance will increase the convergence angle of the electron beam and decrease the spot size. Hence, it results in better image resolution due to the incidence of a stronger electron beam on the specimen surface. Moreover, for a constant electron beam energy, if the working distance is reduced, the spot size decreases and the magnification of the image increases. The effect of working distance on spot size is schematically represented in Figure 3.11.

3.10 EXPERIMENTAL PROCEDURE

The experimental procedure, including specimen preparation, testing conditions and a step-by-step procedure for operating SEM, is explained in this section.

3.10.1 Specimen Preparation

Based on the material to be characterised, different methods are adopted for specimen preparation. The samples/specimens should be free from any dirt or contamination to obtain good-quality images. They should always be kept in a clean environment, such as in desiccators, sealed packets or vacuum chambers. Storing them in a desiccator helps prevent hygroscopic samples from coming into a contact with air and moisture. If the surface of the specimen is coated with oil or grease or exposed to air for a prolonged duration, the layers of contaminants accumulated over the specimen surface should be removed by treating it with solvents and other methods of cleaning before imaging.

The use of wet samples for imaging causes clogging of exits in the microscope as the moisture present in the sample evaporates during the evacuation process of the specimen chamber. Hence, it is important to ensure that all the samples used are completely dry. Oven drying is a suitable method for removing moisture from wet samples. However, some of the samples, such as biological tissues, might lose their structure during oven drying. In such cases, other methods such as critical point drying or freeze drying need to be adopted so that they are dried at a slow rate in a controlled manner.

3.10.1.1 Metals

i. *Sampling and cutting:*

The first step in the specimen preparation of metals is selecting and cutting a section that is to be studied using SEM. Sampling involves identifying a suitable location in the specimen that adequately represents the structure, morphology, and chemical composition of the specimen. The section is cut from the specimen using a cutoff blade made up of silicon carbide, aluminium oxide or diamond. During cutting, lubricants or coolants such as emulsions, low-viscous material or water are used to prevent heating up of the cut section surface and to lubricate the cutoff blade. The recommended surface area of specimen sections is around 10 mm^2, considering the ease of specimen preparation and the allowable specimen size inside the chamber. In some cases, such as those involving the study of failure surfaces, the specimen is imaged without cutting it into thin sections. For instance, the fractured surface of a tensioned steel rebar is imaged using SEM by directly placing it under the electron beam without any specimen preparation.

ii. *Impregnation and polishing:*

To obtain a good contrast and a better detailing of the specimen microstructure, the specimen surface has to be well polished. Before polishing, the specimen has to be impregnated with a suitable thermoset (polyester resins) or thermoplastic (epoxy resins) material. The most commonly used material for impregnation is epoxy resins. Impregnation helps to preserve the structure of the material from getting damaged during polishing as the resins fill the pores, voids and cracks of the material. After impregnation, the resin coating on the surface of the specimen is removed by means of fine-saw cutting. This will expose the specimen's surface for good imaging. Finally, an abrasive such as carborundum, aluminium oxide or silicon carbide is used for coarse polishing of the specimen surface. The specimen is kept on a specimen holder in the polishing machine, and it is ground against abrasive papers with different grit sizes (from coarser to finer) attached to a rotating table. Lubricating oils are applied or water is sprayed over the abrasive paper to avoid heat generation during the grinding process and for a better surface finish. The polishing of the specimen is continued until all the scratch marks have disappeared and a smooth, finished surface is obtained. Ultimately, to attain a fine and scratch-free finish, the surface is finely polished using a diamond spray or paste. The quality of polishing is constantly kept under check by using an OM.

iii. *Etching:*

Etching is an additional process carried out on polished specimens to further enhance the contrast between different features such as grain boundaries, orientations and phases in a micrograph. Chemicals like acid or peroxide solutions are used as etchants to reveal these microstructural features. When the specimen is treated with an etchant, it corrodes the grain boundaries at a faster rate than the grain surfaces, leading to the development of pits and crevices along the boundaries. This results in sharp contrast in the image, revealing even the low-energy regions on the specimen surface. The rate of etching should be regulated to prevent excessive etching of the surface. After adequate etching, an inert solvent such as acetone is used to wash the specimen to avoid further corrosion of the surface.

The polished and etched specimens are cleaned using a solvent such as isopropyl alcohol or ethanol in an ultrasonic bath. Once the specimen surface is cleaned, the specimens are attached to a stub using conductive paint or carbon tape. The main purpose of the conductive paint is to provide a conducting path for electrons to pass to the ground after imaging. The stub is then placed on the specimen holder inside the specimen chamber for imaging. Sputter coating is not required for metallic specimens since they are conductive and the primary electrons can easily interact with the metal without any issue of surface charging.

3.10.1.2 Concrete

The microstructural characterisation of non-conductive materials such as cement paste, mortar or concrete can be performed using SEM after adequate specimen preparation. The preparation of a hardened mortar or concrete specimen involves several steps. The first step is to remove the free water present in the specimen by soaking it in isopropanol. This is done to arrest the hydration of cement at the required curing stage. The specimen is cut into thin sections using a diamond blade cutter to expose the fresh specimen surface.

i. *Epoxy impregnation:*

The cut sections are impregnated with low-viscous epoxy resin and hardener in a cylindrical mould. The epoxy impregnation is done for two reasons: firstly, to support the porous microstructure during grinding, and secondly, to enhance the contrast between hydrated products and unhydrated cement. Without epoxy impregnation, the cutting and grinding processes may damage the surface of the specimen by forming fractures, pits and cracks. Thus, the secondary electron image showing the topographical features will be disturbed. Impregnation should be done under vacuum pressure to remove air from the pores and allow easy infiltration of epoxy into the specimen. The impregnated specimen is kept at room temperature for 24 hours for the epoxy resins to polymerise and harden inside the pores of the specimen.

ii. *Polishing:*

The surface of the specimen should be highly polished and smooth to obtain a BSE image with good contrast. A rough surface texture reduces the contrast and decreases the quality of the image. The epoxy-impregnated specimen is removed from the mould after 24 hours, and the specimen surface is polished using abrasive papers with different grit sizes ranging from 45 to 15 μm. The coarse polishing is continued until the border of the specimen is revealed when observed under grazing light. Further, diamond sprays of particle sizes 6, 3, 1 and 0.25 μm are used for fine polishing the surface to remove the damage caused by grinding. Polishing using 0.25-μm-sized diamond sprays is rarely used as it does not lead to much improvement in the surface texture. The quality of polishing is constantly kept under check by using an OM.

The polishing time and the force applied are slightly increased when the size of the diamond grains is smaller. Also, the polishing time is higher for concrete compared to mortar and paste. The polishing should not be prolonged too much to remove all the resins on the surface since it is important that some quantity of resin remains to stabilise the grains on the surface during polishing. The polished specimens are stored in a desiccator for a few days so that all the contaminants present in the specimen are removed and it is well protected from moisture and CO_2 on the day of testing.

Polishing of concrete specimens at early ages of curing is difficult since the concrete might not have developed sufficient strength and the microstructure will get damaged. Moreover, the use of polished sections for studying the interfacial transition zone (ITZ) in concrete is also difficult owing to the erosion of the ITZ caused by different levels of polishing of the aggregate and paste in the ITZ.

iii. *Sputter coating:*

The polished specimens are coated with a thin layer (15–20 nm) of a suitable conductive material such as carbon, gold, chromium, platinum or palladium. Generally, noble metals are used because of their stability and high SE yield. The most commonly used sputter coating material is gold, owing to its high conductivity and small grain size that allows high-resolution imaging. However, if EDS analysis of a specimen is required, then carbon coating is provided since the X-ray peaks of carbon do not overlap with the peaks of other elements. The carbon-coated specimen will be slightly dark, whereas the gold-coated specimen will have a pinkish tinge on the surface. Sputter coating is applied to non-conductive materials to avoid surface charging and enhance image quality and resolution. The coating material acts as a conducting path for the electrons emitted from the specimen to pass without accumulating on the specimen surface.

The two methods of coating a specimen are ion sputtering and vacuum evaporation. In an ion sputter coater (shown in Figure 3.12), positively charged ions are produced under a high vacuum. Since sputtering is done in vacuum conditions, the specimen has to be completely dry before the application of sputter coating. By creating a very high potential difference between two electrodes, the high-energy ions move towards the target metal (carbon, gold, chromium, platinum or palladium). During the bombardment, energy is transferred to the target metal, causing the release of an ionic

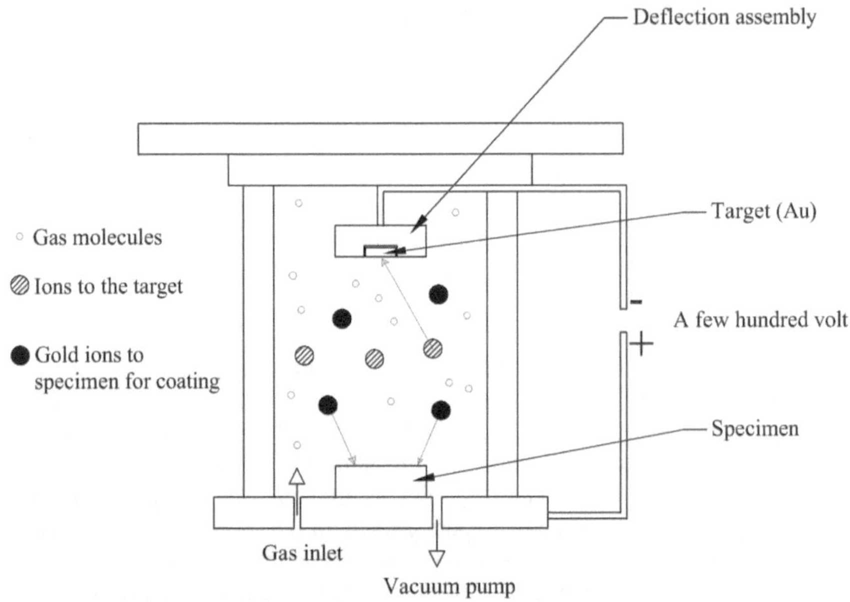

FIGURE 3.12 Ion sputter coater.

cloud at high velocities. The ejected ions collide with some of the residual gas molecules and get scattered in all directions, thus leading to a uniform deposition of the scattered ions on the specimen surface on all sides. Another technique for ion sputtering is using an ion-beam sputter coater. Inside the coater, both the target metal and the specimen are kept under a high vacuum, and a high-quality thin layer of conducting material is formed on the specimen surface. In the vacuum evaporation method of coating, the target material is heated and vaporised under a high vacuum. It then settles on the specimen surface as a thin layer of coating. Since the scattering of the evaporated material particles by colliding with the residual gas molecules is less, the specimen has to be rotated and tilted for the coating to be applied on all sides.

3.10.1.3 Powder Samples

The powder samples are usually sprinkled over a conductive tape made of a material such as carbon. The excess powder on the tape is removed by blowing compressed air with a blower. Care should be taken to ensure that all the powdered particles are well adhered to the conductive tape. This is because loose powder particles will affect the system as they will be sucked out during the evacuation of the specimen chamber and block the evacuation exits of the microscope. The powdered sample can also be impregnated with epoxy, prepared as a flat sample, and polished before imaging. If the particle size is too small (<100 nm), add small quantities of powder in a solvent, ultrasonicate to disperse the particles well, and add a drop of the suspension to a flat surface.

The mortar or concrete specimens can also be ground to powder form instead of being sliced into thin sections for imaging. Since concrete and mortar have cement paste and aggregate, it is important to ensure that both phases are ground to the same level of fineness. A representative sample may be chosen from the interior of the specimen after the concrete cube is crushed during compressive strength testing.

3.10.2 Stepwise Procedure for Operating SEM

The various components of the instrument should be thoroughly checked before operating it. If the instrument is in good working condition, the specimen can be prepared as mentioned in Section 1.10.1. After the specimen is prepared, the following steps are adopted to obtain a micrograph:

Step 1: A double-sided conductive carbon tape is stuck to the stub. Bulk specimens, such as metal, are mounted on the tape, and powdered samples are evenly spread as a thin layer on the tape. It should be ensured that the powder sample is well adhered to the tape and that the particles are not sticking out. Any loose particles should be removed from the surface by using a blower.

Step 2: The specimen exchange in the specimen chamber can be carried out in two ways. In the first method, the entire chamber is vented to atmospheric pressure during the specimen exchange. The chamber is usually vented using dry nitrogen gas to avoid contaminants or moisture in the chamber. In the second method, a specimen pre-evacuation chamber or airlock chamber is used while maintaining a high vacuum in the specimen chamber. The airlock chamber permits specimen exchange without the need to vent the chamber.

Step 3: The stub containing the specimen is placed in a specimen holder and kept inside the specimen chamber. While handling the specimen, tweezers are used to hold the stub in the grooves provided. During the specimen exchange, the valve separating the SEM column and chamber should be closed.

Step 4: After placing the specimen inside the chamber, the vacuum pumps are switched on, and the air is pumped out to re-establish a high vacuum in the chamber.

Step 5: While the vacuum pumping happens, the specimen stage is centred and the working distance is fixed. The movement of the stage is continuously monitored by checking the computer monitor display. The stage height should be carefully set so that the specimen does not collide with the pole piece of the objective lens.

Step 6: The next step is to set the electron beam parameters, such as accelerating voltage and spot size. A high accelerating voltage is adopted for conductive materials, and a low voltage is selected for non-conductive materials to avoid surface charging.

Step 7: Once the chamber has reached the desired vacuum level, the electron gun is switched on and the column valve to the chamber is opened. The electron beam comes down the central axis and strikes the specimen.

Step 8: The scan rate of the electron beam can be changed during beam scanning using a raster scan generator. If the specimen is being moved, then a faster scan rate with a quick screen refresh rate will be used to obtain a live update. On the other hand, when fine focusing is required, a slow scan rate is preferred owing to the high signal-to-noise ratio and better resolution.

Step 9: The initial image shown on the monitor will not be clear. Hence, the image needs to be focused properly. If the image moves while changing the focus, it indicates that the electron gun alignment is not proper. In such a case, the beam alignment should be corrected.

Step 10: Check the image for defects due to astigmatism. If the image is stretched in one direction, it indicates astigmatism caused by the unequal focal length in the two orthogonal directions. The issue should be corrected by adjusting the *x*- and *y*-stigmators. Each time astigmatism correction is applied, the image should be refocused.

Step 11: The magnification is varied to change the scan area and obtain an image with the required details. At higher magnifications, the scan area is smaller and finer focusing needs to be done. A reliable magnification indicator is the calibrated scale bar at the bottom part of the screen displaying the image. This scale bar gives a measure of the size of features on the image.

Step 12: The image is captured after adjusting the contrast and brightness and stored in a database for future reference.

3.11 ENVIRONMENTAL SCANNING ELECTRON MICROSCOPE

Wet specimens such as biological materials and freshly mixed cement paste cannot be studied using SEM owing to the high-vacuum requirement inside the specimen chamber. The microstructure of such specimens changes due to the evaporation of moisture and other volatiles when kept in a vacuum chamber. Moreover, evaporation of moisture from the specimen during the evacuation of the chamber may cause clogging of the exits in the microscope. The variable-pressure SEM and ESEM with a differential pumping system were developed particularly for characterising wet and electrically non-conductive or semi-conductive specimens. The ESEM can be operated at a reduced vacuum pressure of up to 4 kPa.

After the specimen is placed in the specimen chamber, the air inside the chamber is replaced by gases such as water vapour, argon, nitrogen, or carbon dioxide. The ESEM works by slightly increasing the partial pressure of the gas in the chamber. The high pressure is sustained in the specimen chamber by separating it from the ESEM column under a high-vacuum pressure using pressure-limiting apertures. A number of pumping stages are set up in the column to remove the gases that escape from the specimen chamber to the column, thereby maintaining the high-vacuum conditions required for electron guns. Hence, it is possible to examine wet specimens and those containing volatile compounds in their natural condition without being dehydrated. The collision of primary electrons with the gas molecules causes the scattering of the primary electrons, but a fraction still remains unscattered. Instead of the ET detector, a gaseous secondary electron detector based on the gas ionisation principle is most commonly used for collecting SEs in ESEM.

3.12 ENERGY-DISPERSIVE X-RAY SPECTROSCOPY

EDS, or EDX, is an efficient technique adopted for the micro-chemical analysis of a specimen. Combining EDS with SEM helps determine the chemical composition at different locations in the specimen. The technique can be used for qualitative and quantitative analyses of the elements that the material is composed of. It involves the detection of the X-rays with characteristic energy radiated from the specimen during electron bombardment. X-rays with quantum energy in the range of 0.2–20 keV are collected and recorded using an EDS detector fitted in the specimen chamber.

3.12.1 Principle

When an element in a specimen is bombarded with electrons, X-rays of two types are produced. The radiated X-rays are collected, processed by the EDS detector and displayed on the monitor as an X-ray spectrum. A series of X-rays with discrete energy called characteristic X-rays and an X-ray continuum called Bremsstrahlung X-rays together form the X-ray spectrum. In the spectrum, the characteristic lines of elements are superimposed by a continuous X-ray spectrum in the background.

i. *Characteristic X-rays:*

The primary electrons bombarding the specimen interact with the atoms of various elements present in the specimen. The inelastic collision of primary electrons with the tightly bound inner shell (*K* shell) electrons in an atom may lead to the ejection of the excited electrons from the atom, depending on the incident electron energy. For the release of electrons, the incident energy should be greater than the electron-binding energy or critical ionisation energy of the electron. For efficient excitation, the energy of incident electrons should be 3–4 times higher than the critical ionisation energy. Hence, an electron beam of higher energy is required for exciting heavier elements. The total energy imparted to the atomic electron can be up to half the kinetic energy of the incident electron. The resulting vacancy created by the excited electron in the *K* shell is filled by an electron from a higher energy shell (*L*, *M*, etc.). As the electron jumps from a higher energy shell to a lower energy shell, the difference in the binding energy of electrons in the two shells is radiated as an X-ray photon.

The radiation produced during an electron jump from *L* to *K* shell is designated as $K\alpha$ radiation, and that from *M* to *K* shell is designated as $K\beta$ radiation. Similarly, an electron jump from the *M* and *N* shells to the *L* shell results in $L\alpha$ and $L\beta$ radiations, respectively. The emission of these radiations from various atomic shells is schematically represented in Figure 3.13. The energy of the $K\alpha$ radiation (*L* to *K* shell transition) is equal to $E_K–E_L$, where E_K and E_L are the binding energies of electrons in the *K* and *L* shells, respectively. Each shell in an atom has a specific amount of binding energy, also called the atomic energy level. The atomic energy levels in an atom depend on the atomic number of the element. Hence, the energy of X-ray photons released from an atom will be unique to that element.

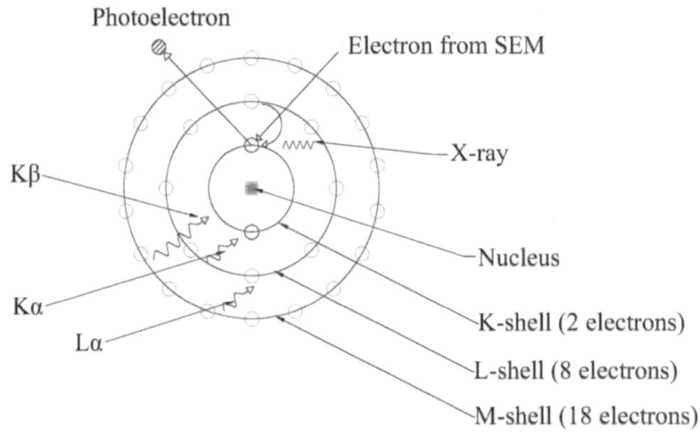

FIGURE 3.13 Characteristic X-ray emission from various energy shells.

ii. *Bremsstrahlung X-rays:*

The Bremsstrahlung X-rays are generated simultaneously with the characteristic X-rays. When the primary electrons travel through the specimen, they experience repulsion from the negatively charged electron cloud in the atoms. As a result, the primary electrons are deflected from their path and lose kinetic energy, which is released as X-ray radiation. Since this interaction occurs randomly, the energy loss of the incident electrons due to the deceleration can range from a few electron volts to the incident electron beam energy. Thus, the Bremsstrahlung X-rays span a wide range of energies and form a continuous X-ray spectrum known as the X-ray continuum.

3.12.2 Factors Affecting X-ray Production

i. *Accelerating voltage:*

The number of X-rays radiated from a specimen and the range of detectable elements are governed by the operating accelerating voltage of the electron gun. A high accelerating voltage is required to provide sufficient energy to excite the inner shell electrons, leading to X-ray photon emission. However, very high accelerating voltage results in lower X-ray spatial resolution and increases the absorption of X-rays due to a larger specimen interaction volume. The typical accelerating voltage used for the X-ray generation from most of the elements is in the range of 15–20 kV.

ii. *Atomic weight:*

The number of X-ray photons produced also depends on the atomic weight of the element. The X-ray emissions are higher when incident electrons strike lighter elements than heavier elements. It is easier to excite *K* shell electrons ($K\alpha$ and $K\beta$ radiations) of elements with lower atomic numbers, whereas significantly high-energy electron beams are required for ejecting the *K* shell electrons of heavy elements. Depending on the incident beam energy, instead of *K* shell electrons, *M* or *L* shell electrons with a comparatively lower binding energy may get excited in the case of the heavy elements. For elements with intermediate atomic weights, both *K* and *L* shell electrons can be released.

3.12.3 Qualitative Analysis

Qualitative analysis involves the identification of elements present in a specimen. The three modes of elemental analysis of a specimen are point analysis, line analysis and area analysis. In point analysis, the X-ray photons radiated from a selected point on the specimen scan area are collected to produce the X-ray spectrum. On the other hand, the distribution of elements along a selected line and a selected area is studied using line analysis and area analysis, respectively. In line analysis, the electron beam is scanned along a line on the specimen to study elemental variation along features such as grain boundaries. X-ray mapping can be done using area analysis. It can be considered an image formed by X-rays generated from the scan area of interest. X-ray mapping helps visualise the dispersion of elements in a specimen.

A plot between intensity in counts or counts per second (*y*-axis) and X-ray energy in keV (*x*-axis) is called an X-ray spectrum. An example of a typical X-ray spectrum is shown in Figure 3.14. The X-rays emitted from different elements have distinct energy positions in the spectrum. Hence, the characteristic X-ray lines in the spectrum are unique for each element. The observed energy of characteristic peaks is compared against the characteristic energies of different elements stored in the computer database, and the peaks are labelled with the element names. The number of X-ray peaks depends on the atomic number of the element. For heavier elements, the number of X-rays generated will be higher, giving rise to a large number of peaks. In order to confirm the presence of an element, it is necessary to identify the majority of these peaks. For the accurate identification of elements, it is essential to ensure that the EDS is properly calibrated using a pure metal such as nickel.

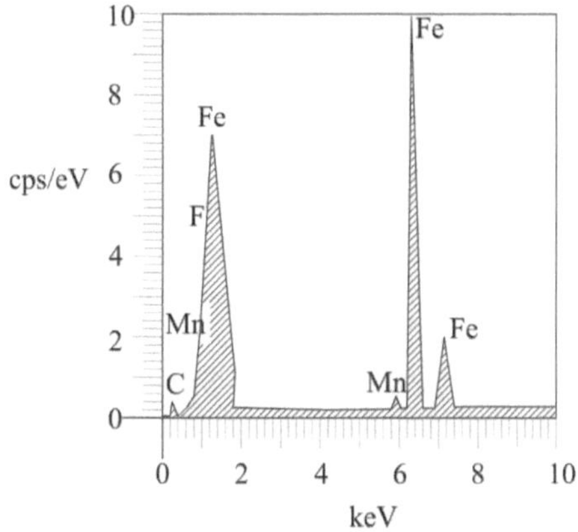

FIGURE 3.14 X-ray spectrum.

The X-ray spectrum is composed of characteristic X-ray lines superimposed with the X-ray continuum, which appears as background noise in the spectrum. The intensity of the X-ray continuum is higher for low-energy electrons and decreases with an increase in incident electron energy. The elements with characteristic energy in the range of the X-ray continuum energy cannot be detected using EDS. To identify a peak, its intensity should be at least three times higher than the intensity of the X-ray continuum present in the background. An electron beam of a smaller size is preferred to obtain a higher peak-to-background ratio owing to its higher energy. Another issue associated with the X-ray spectrum is the overlapping of peaks of certain elements having the same characteristic energy. The appearance of very broad peaks and shoulders in large peaks is attributed to the overlapping of peaks. It is hard to distinguish the peaks in such cases, and the peak stripping feature present in the software should be employed to reveal hidden peaks. The probability of α transitions is greater than β transitions in an atom. Hence, in the X-ray spectrum, the intensity of peaks corresponding to α transitions is higher than that of β transitions.

3.12.4 Quantitative Analysis

The quantitative analysis of elements identified using EDS gives a measure of the concentration of these elements. The intensity of characteristic X-ray peaks is proportional to the concentration of the corresponding element in the specimen. The accuracy with which an element is quantified depends on its concentration in the specimen. The higher the concentration, the greater the accuracy of quantification. Hence, it is easier to quantify major elements (more than 10% by weight of specimen) compared to minor elements present in the specimen. The concentration of an element in a specimen can be obtained by comparing the peak intensity of the element in the unknown specimen with a standard specimen of similar composition. It is important to ensure that the standard specimen is analysed under the same conditions to obtain X-ray spectral data. The quantitative EDS analysis can also be carried out without a standard specimen. In such cases, it is done by comparing the X-ray spectrum of the specimen with the spectral data stored in the computer database. The standard-less analysis is simpler because of the need to analyse only one specimen. The composition of the specimen is expressed in atomic ratios like Ca/Si to minimise errors in quantification.

It is important to note that the electron beam should be incident normal to the specimen surface for a sufficient number of X-ray photon emissions. Further, the specimen should be highly polished to obtain a flat surface without cracks and voids. The voids may scatter the X-rays, and the crests present on a rough surface may block the X-rays from reaching the EDS detector. Hence, proper specimen preparation techniques such as epoxy impregnation and polishing are a must to avoid erroneous measurements. A high-vacuum condition inside the chamber is also a prerequisite for effectively collecting and detecting emitted X-rays. Besides, the distribution of elements in the X-ray generation area should be uniform to accurately quantify the elements. For heterogeneous specimens, the interaction volume includes more than one element, and a mixed X-ray signal is obtained.

3.13 ADVANTAGES OF SEM

The following are the major advantages of using SEM to characterise materials:

i. It is a non-destructive method of analysis.
ii. It can be used to characterise various materials such as metals, ceramics, cement, concrete, polymers and biological specimens.
iii. The material under examination can be conductive or non-conductive, in powder or solid form. Certain materials can be analysed even without any specimen preparation. For example, fractured steel surfaces and corrosion deposits can be imaged by keeping the specimen directly under the electron beam.
iv. The large depth of field of the SEM leads to a larger electron beam-specimen interaction volume. Hence, three-dimensional images of the specimen's microstructure can be obtained.
v. The high image resolution is another key feature of SEM. Modern instruments such as field emission SEM can distinguish points on the specimen surface separated by a distance of only 1 nm.
vi. Rapid imaging makes it a time-efficient method of analysing materials.
vii. Elemental composition analysis can be carried out to detect elements from beryllium to uranium.

3.14 LIMITATIONS OF SEM

SEM technique has the following limitations:

i. The specimen size is restricted to avoid collision of the specimen with the objective lens pole piece as the specimen stage is moved in the vertical direction.
ii. EDS cannot be used for detecting H, He and Li elements. The reason is the lack of a sufficient number of electrons in their atomic shells to release characteristic X-rays when these elements are excited.
iii. The requirement of maintaining a high vacuum inside the SEM column and specimen chamber limits its applicability.
iv. Sputter coating needs to be applied on non-conductive samples to avoid charge accumulation on the specimen surface.

3.15 ISSUES RELATED TO SEM

The imperfections in electromagnetic lenses, such as astigmatism and chromatic and spherical aberration, affect the resolution of EMs. Diffraction at the aperture and surface charging are other common issues that may affect the quality of the micrographs obtained.

3.15.1 Spherical Aberration

Spherical aberration in an electromagnetic lens arises from varying focal lengths in different parts of the lens. The electromagnetic field produced in the lens is stronger

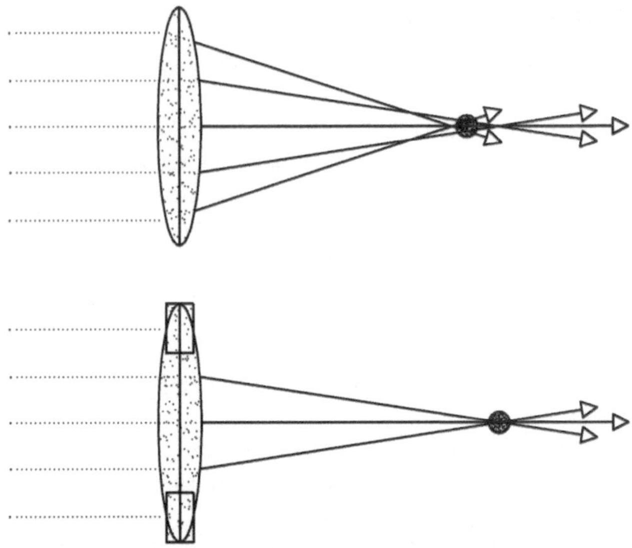

FIGURE 3.15 Spherical aberration and correction.

near the periphery and weaker near the axis of the lens. This causes a change in the convergence angle of the off-axis electrons and results in different focal points. The farther the electrons are from the axis, the more strongly they are deflected towards the axis, causing the focus to shift forward. In such cases, the electrons converge to form a disc instead of focusing on a single point, resulting in a blurred image. To rectify this defect, a small aperture is provided after the objective lens, as shown in Figure 3.15. The aperture blocks the off-axis electrons so that the electrons do not converge at different focal points. However, the provision of an aperture reduces the electron beam diameter and the brightness of the electron beam striking the specimen.

3.15.2 Chromatic Aberration

Chromatic aberration is related to the wavelength or energy of the primary electron beam. It occurs when the electrons emerging from the objective lens are not monochromatic. The difference in the wavelength or energy of the electrons causes them to converge at different angles. The low-energy electrons are bent more towards the central axis, whereas electrons with higher energy converge at a larger distance away from the objective lens. The different focal points result in the formation of a disc rather than a single point of focus. The energy spread of the electron source and the convergence angle of the electron beam have an influence on this defect. For instance, chromatic aberration is more pronounced in a thermionic electron gun, which typically has a larger energy spread than an FE gun. Chromatic aberration can be eliminated by using higher accelerating voltages (>10 keV). However, the variation in electron energy is negligible for most of the electron guns, and hence, in reality, the chromatic aberration will not significantly affect the image resolution.

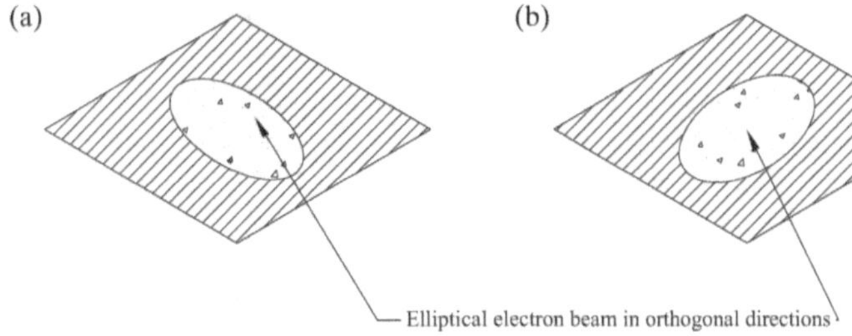

FIGURE 3.16 Astigmatism.

3.15.3 Astigmatism

The ideal, preferred shape of an electron beam is circular and symmetric about the central axis of the SEM column. However, imperfections in the manufacturing of the lenses, accumulation of dirt in the aperture and other problems in the microscope lead to distortion of the beam shape. The non-uniform electromagnetic field produced by a defective lens may lead to the electrons being more focused in one orthogonal direction than the other, as shown in Figure 3.16. The resulting electron beam will be elliptical with two different focal lengths in the orthogonal directions, x and y. This type of electron beam distortion is called astigmatism.

For an electron beam exhibiting astigmatism, the resolution of the image will be degraded in the direction in which the beam cross-section is more elongated. To identify this defect, the electron beam is first under-focused and then over-focused. If the image is sheared and stretched in two opposite directions during under-focusing and over-focusing, it indicates astigmatism. The defect is corrected with the help of the x- and y-stigmator coils and by adjusting the focus to attain a beam of circular shape. The stigmator control knobs and the focus knob provided in the control system are adjusted at a high magnification of about 10,000× until a sharp image is obtained.

3.15.4 Diffraction at Aperture

The electrons may be subjected to diffraction at the edges of the apertures placed in the objective lens assembly. This defect can be reduced by increasing the beam convergence angle, contrary to the smaller convergence angle requirement for avoiding spherical and chromatic aberrations. Hence, an optimum convergence angle should be determined to limit all three lens defects.

3.15.5 Surface Charging

Surface charging occurs due to the difference in incident beam current entering the specimen and total outgoing current leaving the specimen, including secondary electron and BSE current. Not all the electrons striking the specimen leave the specimen as secondary and BSEs; instead, some of the electrons may accumulate within the specimen. The grounding of the specimen stage is essential to avoid charge accumulation on the specimen surface. Grounding provides a continuous conductive path for the electrons to pass from the specimen surface to the ground through the specimen stage.

However, when a non-conductive material is placed inside the chamber for imaging, electrons striking the specimen surface will not reach the grounded specimen stage due to the absence of a conducting path. As a result, the electrons get collected at local points on the specimen surface. The accumulation of negative charges on the localised specimen surface is known as specimen charging. The resulting charge buildup will affect the emission of SEs from the specimen. This effect can be observed in the SE image in many ways, such as highly bright regions in the image, scan discontinuities/horizontal lines in the image, image distortion, etc. A point to be noted is that the charging effect is more pronounced at higher magnifications. Non-conductive materials are sputter coated with a thin film of conductive material like carbon, gold, platinum or palladium to prevent surface charging. Double-sided conductive tapes made of carbon are used to stick the specimen to the stub for easy conductance of electrons. To further reduce the charging effect, the following measures can be adopted:

i. Reduce the electron beam current
ii. Lower the accelerating voltage
iii. Adjust the orientation of the specimen by rotating or tilting it
iv. Adopt a quick scan rate and frame integration

3.16 APPLICATION OF SEM

Image analysis of specimen microstructure using a combination of elemental maps and micrographs in SE and BSE modes is a reliable characterisation technique.

3.16.1 Phase Identification

BSE images of materials can be used for phase identification owing to the compositional contrast in the images. Compositional contrast is observed due to differences in the intensity of signals generated from regions or phases with different elemental compositions. This intensity variation is proportional to the difference in the average atomic weight of the respective regions in the specimen. A region with a higher atomic weight produces a greater number of BSEs and appears brighter relative to a region with a lower atomic weight on a BSE image. The contrast will be greater for elements with a large difference in their atomic numbers.

BSE images can also be used for quantitative phase analysis of materials. The volume fraction of a phase in a material is equal to the average surface fraction in a micrograph in BSE mode. A significant number of BSE images of well-polished sections that are representative of the material need to be recorded to obtain accurate results. The number of images required depends on the heterogeneity of the material.

3.16.2 Topographical Microstructure

The SE mode is the most widely used imaging mode of SEM to study the shape and morphology of various phases in a material. The SE mode gives information regarding the topographical microstructure by processing the secondary electrons generated from the specimen. The SE images have a three-dimensional appearance due to depth of field limitations.

3.16.3 Application in Concrete Technology

The microstructure of concrete is heterogeneous and complex. Unlike other engineering materials, the microstructure of concrete is subject to change with time and environmental conditions. A good knowledge of the microstructure of concrete is essential to understanding its long-term performance. SEM combined with microanalysis using EDS has paved the way to explore and understand the various mechanisms in concrete at the micro-level.

Some of the main applications of SEM in concrete technology are:

i. Study the microstructure of clinker and hydrated phases in cementitious systems.
ii. Study the distribution of phases in the cement matrix.
iii. Quantitative analysis of cement and hydrated cement products.
iv. Study the morphology of pozzolanic materials.
v. Determine the porosity and pore connectivity of the paste matrix and aggregate it in concrete.
vi. Study the cause, extent and mechanism of deterioration of concrete due to external chemical attacks, alkali–silica reactions (ASRs), delayed ettringite formation, etc.
vii. Determine the chemical composition of corrosion products formed on corroded steel surfaces and identify the corrosion mechanism.

If the water-to-cement ratio is known, the degree of hydration can be determined by measuring the surface fraction of unreacted cement particles in the BSE image of the hydrated cement. For a cement paste, 50–100 BSE images are required, whereas for a mortar and a concrete specimen, more than 100 images are required to obtain representative results of quantitative phase analysis. The amount of portlandite present in the specimen cannot be estimated accurately since it is present as small clusters in the microstructure, leading to an underestimation or overestimation of its quantity. Thermogravimetric analysis is more accurate in determining the quantity of portlandite and carbonates in cementitious systems compared to SEM. Similarly, Rietveld analysis using XRD data is preferred for the quantification of crystalline phases such as alite and belite present in cement.

3.16.3.1 Phases Typically Observed in Cementitious Systems

i. *Cement paste:*

The microstructure of a hydrated cement paste consists of C–S–H, portlandite, and calcium sulfoaluminate phases. Partially hydrated and unhydrated cement grains can also be easily identified in the micrograph of a hydrated cement paste. The unhydrated cement grains appear as extremely bright spots in the BSE image. In addition to solid phases, the paste matrix contains different types of voids or pores. The capillary pores are seen as dark areas, whereas the gel pores are too small to be resolved by SEM. All these features are easily distinguished in a BSE image based on the contrast level or an SE image based on the morphology. When cement paste specimens with the same mix proportion are examined in BSE mode at different curing ages, it is observed that the volume of voids in the paste matrix reduces as the age of hydration increases.

C–S–H is the most important phase in concrete since it has binding properties and provides strength to hardened concrete. It occupies 50%–60% of the total solid volume in the cement paste. The C–S–H phase is poorly crystalline and has an unresolved morphology. The structure of C–S–H varies from poorly crystalline fibres to reticular networks. The outer- or early-product C–S–H is observed in the pores where capillary water was initially present. Once the hydration of cement becomes diffusion-controlled, a rim of inner product C–S–H is formed around the hydrated cement grains. The inner product C–S–H can be easily distinguished from the former as it is denser. Secondly, calcium hydroxide (CH), also known as portlandite, makes up 20%–25% of the solid volume in the hydrated paste. They are purely crystalline and appear in different shapes and sizes. CH phases are mostly large, platy crystals with hexagonal prism morphology or large, elongated crystals. Due to their large sizes, they can be easily recognised in the micrograph of a cement paste.

The calcium sulfoaluminate phases (ettringite and monosulfoaluminates) occupy 10%–15% of the solid volume in the hydrated paste. During early curing days, the ettringite is formed in the voids present in the cement matrix. The ettringites produced during early hydration are long and slender, needle-shaped crystals. When ettringite is formed at a later stage, it appears as agglomerates of needle-shaped crystals that absorb water and expand. On the other hand, the monosulfoaluminates are formed at later stages of cement hydration, depending on the alumina-to-sulphate ratio in the cement. Monosulfoaluminates are found in the hydrated cement paste with two different morphologies. Initially, the monosulfoaluminates form clusters or rosettes of irregular plates, and later, they develop into very thin hexagonal plates. The hexagonal plate structures are also characteristic of calcium aluminate hydrates that are produced when tricalcium aluminate (C_3A) reacts with water in the absence of gypsum.

ii. *Aggregates:*

The aggregates in concrete are mainly responsible for the elastic modulus, dimensional stability, and unit weight of concrete. Therefore, the physical characteristics of the aggregates, such as volume, size, texture, shape and distribution of pores, are more significant than the chemical characteristics of the aggregates. SEM is useful for determining the size, shape, texture and porosity of aggregates. Moreover, the chemical composition of deleterious constituents present in aggregates can be identified using EDS.

iii. *Interfacial transition zone:*

The aggregate surfaces are generally covered by a thin layer of hydration products such as CH and C–S–H. Beyond this thin film, the ITZ is observed. The ITZ is described as the weakest zone in concrete since it is highly porous compared to cement paste and aggregate. SE images of concrete can be used to study the microstructural features present in the ITZs in concrete. The SE image focusing on an ITZ shows numerous platy portlandite crystals and ettringite. Moreover, the cement particles in the ITZ are often fully hydrated, indicating a high water-to-cement ratio at the interfacial zone.

3.16.3.2 Deterioration in Concrete due to Chemical Attack

The reason and extent of deterioration in concrete can be determined by studying the microstructure of concrete using SEM. The deterioration may be due to an external chemical attack such as carbonation, external sulphate attack and corrosion, or an internal chemical attack such as ASR and delayed ettringite formation. Freeze-thaw cycles, fire, weathering, and crack formation due to loading are some of the physical causes of deterioration in concrete. The SEM combined with EDS analysis provides a great insight into the factors that influence the durability properties of concrete, such as the microstructure of the hydrated paste, the permeability of the concrete, matrix densification, chemical composition of the phases, and the structure of the ITZ. The main products of a chemical attack on concrete can also be identified using SEM/EDS analysis.

Carbonation of concrete occurs when atmospheric CO_2 reacts with portlandite to produce calcite. The calcite crystals have different morphologies, such as granular, prismatic, globular and dogtooth shapes. At a later stage, the CO_2 also reacts with C–S–H causing the decalcification of C–S–H. The disintegration of C–S–H leads to the development of a porous microstructure, as observed in the micrograph of carbonated concrete. On the other hand, when sulphates from external sources such as soil, groundwater or seawater diffuse into concrete, they react with hydrated cement products to form gypsum. The gypsum formed reacts with unhydrated C_3A and monosulfoaluminates to produce secondary ettringite. Ettringite, being expansive in nature, will absorb pore water present in concrete, which causes cracking in concrete due to the expansive pressure developed. Sometimes, another phase known as thaumasite is observed in the micrograph of concrete damaged by a sulphate attack. Thaumasite is formed at low temperatures when carbonate and sulphate ions are available. The structure of thaumasite is similar to the needlelike structure of ettringite. EDS analysis is used to distinguish thaumasite and ettringite phases in a concrete specimen.

ASR occurs when the aggregates used in concrete are composed of reactive silica. The alkalis in the pore solution react with the reactive silica in the aggregates to form an ASR gel around the aggregates. The ASR gel is hygroscopic and expansive in nature, leading to the development of expansive pressure and resulting in cracking in concrete. The BSE images of the specimen can be used to distinguish the aggregate, cement matrix, and ASR-damaged areas based on the contrast.

3.16.3.3 Microstructure of Pozzolanic Materials and Blended Concrete

Pozzolanic materials such as fly ash, silica fume, rice husk ash (RHA) and sugarcane bagasse ash (SCBA) are used as supplementary cementitious materials (SCM) owing to their high amorphous silica content and pozzolanic reactivity. The pozzolanic materials react with portlandite present in hydrated cement to produce secondary C–S–H, thereby enhancing the strength of concrete blended with pozzolanic materials. In general, the addition of pozzolanic materials to concrete changes its microstructure and results in high-performance concrete. The pozzolanic reaction of these materials leads to a reduction in the amount of CH and an increase in the amount of C–S–H formed. The micrograph of blended concrete shows a denser microstructure and less porous ITZ. The higher impermeability of blended concrete leads to better resistance to chemical intrusion and, thereby, enhances the durability properties of concrete. The microanalysis of the C–S–H produced by the pozzolanic reaction showed that it has a similar composition as the C–S–H produced by cement hydration, but the Ca/Si ratio is lower. Calcium aluminosilicate hydrate (C–A–S–H) is an additional hydration product observed when SCMs rich in aluminates are blended with cement. Similar to C–S–H produced from cement hydration, C–A–S–H also has binding properties and results in a denser microstructure in blended concrete.

Fly ash is a by-product of thermal power plants that results from the combustion of coal. The microstructure of fly ash consists of spherical particles of various types and sizes. The fly ash particles may be solid spheres, cenospheres, which are hollow spheres, or plerospheres, which are spheres filled with smaller spheres. Ground-granulated blast furnace slag is another industrial by-product that is popularly used as an SCM. It is a by-product obtained during the extraction of iron from iron ores in the blast furnace of the steel industry that is rapidly cooled and ground to a fine size. The SE image of slag shows irregularly shaped particles of a size larger than cement particles. In addition to the pozzolanic behaviour, slag can directly react with water to form C–S–H at a slow rate, which is known as the latent hydraulic property of slag. Hence, slag is considered both cementitious and pozzolanic in nature. This can be confirmed by the presence of a higher amount of C–S–H and a lower amount of CH in the microstructure of slag-blended concrete compared to OPC concrete. The hydration products of slag can be observed in the micrograph as a rim around the slag particles, and the size of the rim increases as the hydration progresses. In addition to being an excellent pozzolanic material, silica fume acts as a filler due to its ultra-fine spherical particles. The finer particles of silica fume fill the pores present in the microstructure and reduce the porosity of concrete. Therefore, the microstructure of silica fume-bleached concrete is even denser than other blended concrete.

Even though most of the agro-waste ashes are rich in amorphous silica, the large particle size and presence of large amounts of unburnt carbon in the as-received ashes hinder their acceptance as an effective substitute for cement. Figures 3.17a, e and g show the micrographs of as-received SCBA, RHA and palm oil fuel ash (POFA), respectively. The as-received agro-waste ashes were found to have completely burned particles and large-sized fibrous unburnt particles. Various processing methods such as sieving, grinding, burning and chemical treatment can be adopted

a. As-received SCBA
b. Sieved and ground SCBA
c. SCBA burnt at 500 °C
d. SCBA burnt at 900 °C
e. As-received RHA
f. Sieved and ground RHA
g. As-received POFA
h. Sieved and ground POFA

FIGURE 3.17 Micrographs of as-received and processed agro-waste ashes.

to eliminate the unburnt carbon and enhance the pozzolanic reactivity of agro-waste ashes. The method and optimum conditions of processing should be judiciously chosen depending on the type, source and required pozzolanic reactivity of the ashes.

Sieving and grinding can be considered an effective method to remove most of the coarse and fibrous particles present in the as-received SCBA and POFA, as sieved and ground ashes were characterised by finer particles and the absence of fibrous particles (as shown in Figure 3.17b and h). The EDS analysis of fibrous particles present in the as-received SCBA has revealed that they are mainly composed of carbon, as depicted in Figure 3.18a. Moreover, the prismatic and angular particles in SCBA were found to be rich in silica (shown in Figure 3.18b). Similarly, the highly porous cellular microstructure of as-received RHA has disappeared, and irregularly shaped

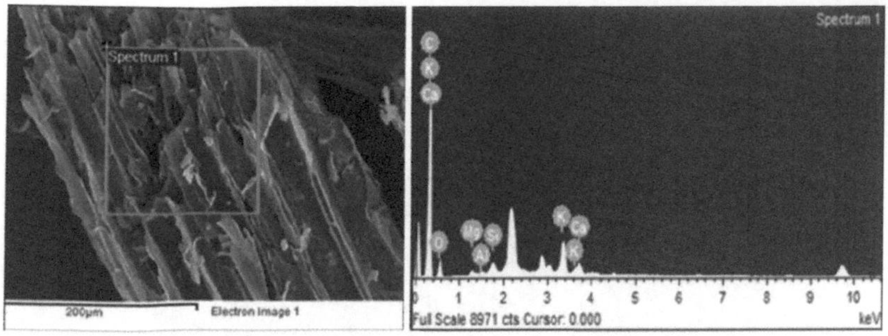

a. Fibrous particle

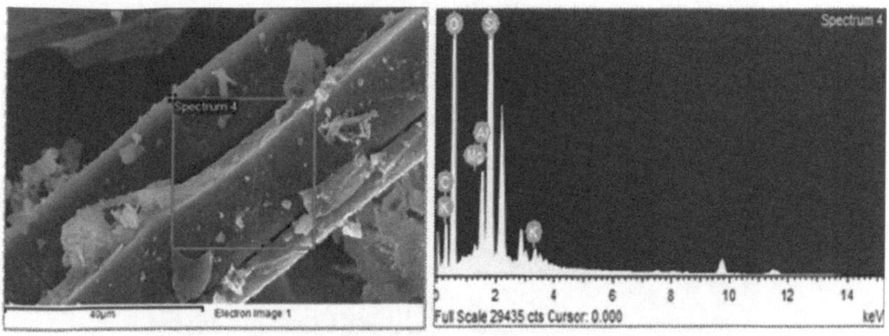

b. Prismatic particle

FIGURE 3.18 Energy-dispersive X-ray spectrometer analysis of fibrous and prismatic particles in sugarcane bagasse ash.

particles of finer size are visible in sieved and ground RHA (shown in Figure 3.17f). The breakdown of cellular structure helps to overcome the problems associated with water absorption and workability of RHA-blended concrete. The effect of burning temperature can also be studied by examining the microstructure of the ashes burned at various temperatures. For instance, as-received SCBA was mainly composed of large fibrous particles, whereas the micrograph of SCBA burned at 500°C (shown in Figure 3.17c) shows various kinds of particles, such as prismatic, spherical, fibrous and irregular particles. Moreover, Figure 3.17d, showing the microstructure of SCBA burned at 900°C, reveals the presence of coarser particles, indicating the conversion of amorphous silica to crystalline silica. Therefore, understanding the microstructural characteristics of agro-waste ashes is important to determine the degree of processing required and assess the influence of various processing methods on the quality of ashes.

3.16.4 Application in Other Civil Engineering Fields

3.16.4.1 Pavement Materials

- The microstructure of the asphalt binder can be studied using SEM. A suitable preparation method should be adopted for the raw sample before examining it under the electron beam. This is because of the difficulty of scanning non-conductive oily samples, such as asphalt, using SEM. Hence, the sample is pre-treated using a solvent to eliminate the oil phase while preserving the asphaltene structure. The micrograph of asphaltene is typically a network structure with interconnected spherical particles.

- SEM is a useful technique for the characterisation of aggregate morphology. The micrograph of aggregates shows different morphologies such as flaky particles, spherical particles, flocculent particles and irregular stripes. The texture of an aggregate surface can also be studied using SEM. Moreover, EDS can be used to determine the elemental composition of aggregates, which is helpful to further determine the mineral composition using XRD.
- The performance of asphalt pavement depends on the strength of the asphalt binder, the aggregate and the asphalt–aggregate interface. The microstructure of the asphalt–aggregate interface can be studied using SEM to better understand the adhesion or binding between the two phases. For instance, the microstructure of the asphalt-limestone aggregate interface can be examined to determine the penetration depth of asphalt into the pits and fractures present on the surface of the limestone. The penetration of asphalt into the aggregate surface will enhance the strength of the interface. On the other hand, granite aggregate has a smooth surface, leading to an adhered asphalt layer on the aggregate surface and no penetration of asphalt into the aggregate.

3.16.4.2 Soil

- The size, shape and distribution of particles, connectivity and distribution of pores in the soil sample are a few parameters that can be determined using SEM. Another electron microscopy technique used for the microstructural characterisation of soil is ESEM. The main advantage of ESEM is that wet soil samples can be studied, thereby preserving their natural state for examination.
- The micrograph of clayey soil exhibits various morphologies depending on the origin of the soil. Honeycomb, skeletal, matrix, turbulent and laminar microstructural types are typically observed in clay of sedimentary origin. In contrast, clayey soils of eluvial and hydrothermal origin are dominated by pseudoglobular, domain and sponge structures.
- The EDS analysis can be used to determine the elemental composition of the soil. In general, significant quantities of black carbon produced by incomplete combustion of biomass, coal, or oil are present in the soil. Based on the EDS analysis, the black carbon and organic matter present in the soil can be appropriately distinguished. The source and properties of black carbon can be determined by knowing its morphology and chemical characteristics. The black carbon particles present in soil may have various morphologies, ranging from spherical to irregular shapes. For example, charcoal particles are larger (μm to m) and have a fibrous texture. In contrast, soot particles are smaller (submicrometer scale) and have a spherical shape.
- The swelling-shrinking behaviour of soil at a microstructural scale can be investigated using ESEM combined with digital image analysis.

BIBLIOGRAPHY

Akhtar, K., Khan, S. A., Khan, S. B., & Asiri, A. M. (2018). "Scanning electron microscopy: Principle and applications in nanomaterials characterization". In: S. Sharma (Eds.) *Handbook of Materials Characterization* (pp. 113–145). Springer, Cham, Switzerland. https://doi.org/10.1007/978-3-319-92955-2_4

Goldstein, J., & Yakowitz H. (2012). *Practical Scanning Electron Microscopy: Electron and Ion Microprobe Analysis*. Springer, New York. https://doi.org/10.1007/978-1-4613-4422-3

Inkson, B. J. (2016). Scanning electron microscopy (SEM) and transmission electron microscopy (TEM) for materials characterization. In *Materials Characterization Using Nondestructive Evaluation (NDE) Methods* (pp. 17–43). Woodhead Publishing, UK. https://doi.org/10.1016/B978-0-08-100040-3.00002-X

Reimer, L. (2000). *Scanning Electron Microscopy: Physics of Image Formation and Microanalysis. Springer Series in Optical Sciences* (Vol. 45). Springer-Verlag, Berlin, Heidelberg. https://doi.org/10.1007/978-3-540-38967-5

Ul-Hamid, A. (2018). *A Beginners' Guide to Scanning Electron Microscopy* (Vol. 1, p. 402). Springer, Cham, Switzerland. https://doi.org/10.1007/978-3-319-98482-7

Watt, I. M. (1997). *The Principles and Practice of Electron Microscopy*. Cambridge University Press, England. https://doi.org/10.1017/CBO9781139170529

MULTIPLE CHOICE QUESTIONS

1. Which of the following statements related to electron microscopy are true?
   a. Electrons generated by an electron gun have a shorter wavelength than light waves produced by a tungsten-halogen lamp.
   b. The resolution of an electron microscope increases as the wavelength of the electron beam increases.
   c. Electron beam emerging from electron lenses are divergent, and divergence slits are placed in its path to produce a collimated beam.
   d. Electron microscopes are operated at atmospheric pressure.
2. Higher accelerating voltage in the electron gun causes _______.
   a. Higher image resolution
   b. Surface charging of the specimen surface
   c. Damage to low-conductive materials
   d. All of the above
3. Define working distance in SEM.
   a. Distance between scan coils and specimen surface.
   b. Distance between the pole piece of the objective lens and the specimen surface.
   c. Depth of the specimen for which all the features are in good focus.
   d. Distance from condenser lens to objective lens.
4. Which of the following factors influences the specimen interaction volume of the incident electrons?
   i. Elemental composition of the specimen
   ii. Accelerating voltage in the electron gun
   iii. Angle of incidence of electron beam
   iv. Intensity of electron beam

a. i and ii
b. i, ii and iii
c. ii, iii and iv
d. i, ii, iii and iv

5. **Statement A:** The penetration depth and scattering of incident electrons are lower for a denser specimen.
Statement B: The absorption of incident electrons is greater than scattering by a denser specimen.
a. Statement A is true, and Statement B is false
b. Statement A is false, and Statement B is true
c. Both the statements are correct, and B is the correct explanation for A
d. Both the statements are correct, but B is not the correct explanation for A

6. What are auger electrons?
a. When high-energy primary electrons overcome the binding energy of tightly bound inner shell electrons, secondary electrons known as auger electrons are ejected.
b. The electrons scattered by positively charged nucleus of an atom are known as auger electrons.
c. The de-excitation energy released during electron transition is absorbed by another electron in the atom, leading to ejection of the electron. The ejected electron is called an auger electron.
d. When high-energy primary electrons overcome the binding energy of loosely bound inner shell electrons, secondary electrons known as auger electrons are ejected.

7. Which of the following statements are true about secondary electrons?
a. The energy of secondary electrons is lower than 50% of the primary electrons.
b. Polishing and grinding the specimen surface is not required for secondary electron microscopy.
c. The SE images represent the topographical features of a specimen.
d. All of the above.

8. What are the different types of secondary electron signals?
i. Electrons produced during the inelastic collision of the incident electrons with the loosely bound valence electrons in the atom.
ii. Electrons produced during the elastic collision of the incident electrons with nucleus of an atom.
iii. Electrons produced during inelastic scattering of backscattered electrons with other secondary electrons.
iv. Electrons generated when the backscattered electrons strike the chamber walls or objective lens system of the instrument.
a. i, iii and iv
b. i, ii, iii and iv
c. ii, iii and iv
d. i and ii

9. Which of the following are applications of the backscattered electron imaging?
   a. Size and orientation of grains in the microstructure of a specimen.
   b. Identify voids, cracks and fissures in a specimen.
   c. Morphology of various phases in a specimen.
   d. Both a and b.
10. What do you mean by cathodoluminescence?
    a. The emission of fluorescent X-rays when electrons strike a luminescent material.
    b. The emission of light photons when electrons strike a luminescent material.
    c. The emission of X-rays when light photons strike a luminescent material.
    d. None of the above.
11. Why is SEM column always kept under high vacuum?
    a. To prevent oxidation and burning out of tungsten filament at high temperatures.
    b. To prevent the interaction between the gases in the column and the electron beam which may produce random discharge.
    c. To provide a clean and dust-free environment.
    d. All of the above.
12. What is the function of a Wehnelt cap in an electron gun?
    a. To accelerate the electrons produced from the cathode towards the specimen.
    b. To protect the anode and cathode from getting contaminated by foreign materials.
    c. To ensure that the electron beam doesn't get scattered and stays focused on the central path.
    d. To prevent overheating of filament.
13. Which of the following statements are true about thermionic and field emission electron guns?
    a. Thermionic electron gun has lower brightness and shorter lifetime than FE electron gun.
    b. The diameter of the electron beam produced by an FE electron gun is smaller resulting in higher-resolution images.
    c. FE electron gun has a lower energy spread than thermionic electron gun.
    d. All of the above.
14. What is the disadvantage of a slow scan rate?
    a. Lower signal-to-noise ratio
    b. Blurry images due to image drift
    c. Low resolution
    d. Surface charging

15. What is the function of stigmator coils?
    a. Scans the electron beam over the specimen surface.
    b. Reduces the beam size and focus the electron beam on the specimen surface.
    c. Corrects distorted shape of an electron beam to produce a circular beam.
    d. Prevents the stray electrons from reaching the objective lens.
16. **Statement A:** The Everhart–Thornley detector can effectively separate secondary electrons from the backscattered electrons.
    Statement B: The ET detector is placed above the objective lens and the magnetic field produced by the objective lens is used to capture the electrons.
    a. Statement A is true, and Statement B is false
    b. Statement A is false, and Statement B is true
    c. Both the statements are true
    d. Both the statements are false
17. At what position is the collection efficiency of BSE detector maximum?
    a. When it is placed perpendicular to the specimen surface.
    b. When it is placed at an angle between 0° and 90° from the specimen surface.
    c. When it is placed parallel to the specimen surface.
    d. Collection efficiency is independent of the detector position.
18. Which of the following elements cannot be detected using an EDS detector?
    a. Noble gases
    b. H, He and Li
    c. Alkali metals
    d. None of these
19. What is the function of the thin opaque window provided inside the EDS detector?
    a. Protects the semiconductor crystal from visible light radiation.
    b. Helps maintain vacuum inside the detector.
    c. Attracts the X-rays into the collimated tube.
    d. Both a and b.
20. What is the need for epoxy impregnation of specimens?
    a. To provide a smooth polished surface.
    b. To prevent the contamination of the specimen by dust, chemicals, etc.
    c. To preserve the microstructure during polishing of the specimen.
    d. To enhance the strength of secondary electron signal generated.
21. Which of the following statements are true about polishing a concrete or mortar specimen?
    a. Coarse polishing is done using a carborundum paper and diamond sprays are used for fine polishing.
    b. The polishing time is higher for mortar compared to concrete.
    c. Polished specimens are preferred for studying ITZ in concrete.
    d. All of the above.

22. Why is sputter coating applied to a specimen before imaging?
    a. To obtain a backscattered electron image of the specimen with good contrast.
    b. To avoid surface charging of a non-conducting specimen.
    c. To preserve the microstructure during polishing of the specimen.
    d. To prevent the contamination of the specimen by dust, chemicals, etc.
23. What is edge effect?
    a. When the electron beam strikes the specimen holder instead of the specimen, it appears as a bright spot in the micrograph.
    b. When the incident electrons are accumulated on the specimen surface due to the absence of a non-conductive path, it appears as a bright spot in the micrograph.
    c. When the electron beam strikes inclined portions of the specimen surface such as ridges, edges and slopes, they appear as bright spots in the micrograph.
    d. None of the above.
24. Increase in the accelerating voltage results in ________.
    a. Increase in electron beam current
    b. Decrease in electron beam size
    c. Better image resolution
    d. All of the above
25. Which of the following are true related to chemical analysis using EDS?
    a. Can be used for both quantitative and qualitative elemental analysis of specimens.
    b. Elements of higher atomic number produce lesser number of peaks in the X-ray spectrum.
    c. Concentration of elements cannot be determined using EDS analysis.
    d. All of the above.
26. Which of the following are the main advantages of using SEM for characterising materials?
    i. Conductive and non-conductive materials can be examined using SEM.
    ii. Fractured surface of specimen can be examined by directly placing the specimen under the microscope.
    iii. All the elements in the periodic table can be identified using SEM-EDS analysis.
    iv. Unlike XRD, high vacuum is not required inside the instrument while conducting the experiment.
    a. i, iii and iv
    b. i, ii, and iii
    c. i and ii
    d. i, ii, iii and iv

27. How is the issue of spherical aberration of objective lens resolved?
    a. By placing an aperture after the objective lens to block the off-axis electrons.
    b. By placing electromagnets after the objective lens to refocus the off-axis electrons.
    c. By adjusting the electromagnetic field strength produced by the objective lens.
    d. All of the above.
28. Explain chromatic aberration in electron microscopes.
    a. The EM field produced in the lens is stronger near the periphery of the lens, due to which off-axis electrons are converged at different focal points.
    b. Imperfections in the lens, and contamination of lens and aperture by dirt cause distortion of the electron beam shape.
    c. Polychromatic electron beam leads to disc shaped focus instead of a single focal point.
    d. Imperfections in the lens produce an electron beam of polychromatic nature.
29. Which of the following steps can be adopted to reduce surface charging of the specimen?
    i. Coat the specimen surface with conductive material such as gold, palladium and platinum.
    ii. Increase the electron beam current.
    iii. Adopt a quick scan rate and frame integration.
    iv. Increase the accelerating voltage of the electron beam.
    v. Adjust the orientation of the specimen.
    a. i, ii and v
    b. i, ii, iii and v
    c. i, iii, iv and v
    d. i, iii and v
30. Which of the following are the applications of SEM in concrete technology?
    a. Study the morphology of phases present in cementitious systems.
    b. Study the cause and extent of deterioration of concrete.
    c. Determine the degree of hydration in concrete specimen.
    d. All of the above.
31. Among the following, identify the incorrect statements related to the microstructure of concrete?
    i. Unhydrated cement particles appear as dark spots and pores appear as bright spots in the BSE image.
    ii. Unhydrated cement particles appear as bright spots and pores appear as dark spots in the BSE image.
    iii. Portlandite is observed as hexagonal crystal, and ettringite as needle-shaped crystal in SE image.
    iv. Numerous CH and ettringite crystals are observed in the interfacial transition zone of concrete.

a. i, and iv
b. ii, and iv
c. only i
d. i, and iii

32. Thaumasite crystals are observed in the microstructure of concrete due to which of the following reasons?
    a. Ingression of H_2SO_4 into concrete.
    b. Presence of carbonate and sulphate ions in concrete.
    c. Ingression of $MgSO_4$ or Na_2SO_4 into concrete.
    d. Diffusion of CO_2 from atmosphere into concrete.
33. What are the changes observed in the microstructure of concrete when cement is blended with pozzolanic materials?
    a. Increase in the number of portlandite crystals and C–S–H.
    b. Decrease in the number of portlandite crystals and increase in the amount of C–S–H.
    c. Denser microstructure.
    d. Both b and c.

1 a	2 d	3 b	4 b	5 c	6 c	7 d	8 a	9 d	10b
11d	12c	13d	14b	15c	16d	17a	18b	19d	20c
21a	22b	23c	24d	25a	26c	27a	28c	29d	30d
31c	32b	33d							

4 Thermogravimetric Analysis

4.1 OVERVIEW

Thermal analysis, in general, is the study of material properties such as weight loss as a function of temperature. Absorption, adsorption, vaporisation, sublimation, decomposition, oxidation, reduction, and crystallisation of materials are some of the phenomena that are influenced by temperature variation. In addition, volatile or gaseous products present in the materials or those formed due to the thermal reaction can be analysed. Thermal analysis can also aid in assessing the thermal stability of a material.

The present chapter gives a brief description of the terminologies related to thermal analysis and the instrumentation of a thermogravimetric analyser. The factors influencing the measurement using thermogravimetric analysis (TGA), including specimen characteristics and instrument factors, are also discussed in detail. An in-depth explanation of the working principle of TGA and a stepwise procedure for conducting the experiment are given in the subsequent sections. A section focusing on the application of TGA in cement chemistry is present, which provides aid for the thorough understanding of phase identification and quantification of cement hydrates. Differential scanning calorimetry (DSC) and differential thermal analysis (DTA) methods of analysis are also briefly described in this chapter.

4.2 TYPES OF THERMAL ANALYSIS

Figure 4.1 shows the classification of thermal analysis and the main parameters measured in each method. In TGA, the variation in the weight of the sample is determined as a function of temperature and/or time. All phenomena that are accompanied by weight loss, such as decomposition, vaporisation, absorption and crystallisation, can be studied using TGA. DSC and DTA are other thermal analysis techniques often used for the characterisation of materials. DSC measures the enthalpy change, whereas DTA measures the temperature difference between the sample and the reference material. The evolved gas analysis is used to determine the nature and amount of gaseous or volatile components evolved from a material undergoing thermal analysis. It is often combined with other techniques such as TGA, Fourier transform infrared spectroscopy, gas chromatography and mass spectroscopy. This technique is useful to determine the concentration and composition of evolved gases. Thermomechanical analysis and dynamic mechanical analysis are used to measure mechanical properties such as deformations and dynamic moduli. In thermomechanical analysis, a static load is applied to the sample, and the deformation with a change in temperature or time is measured. On the other hand, in dynamic mechanical analysis, a sinusoidal load is applied to study the viscoelastic properties of a material.

DOI: 10.1201/9781032635392-4

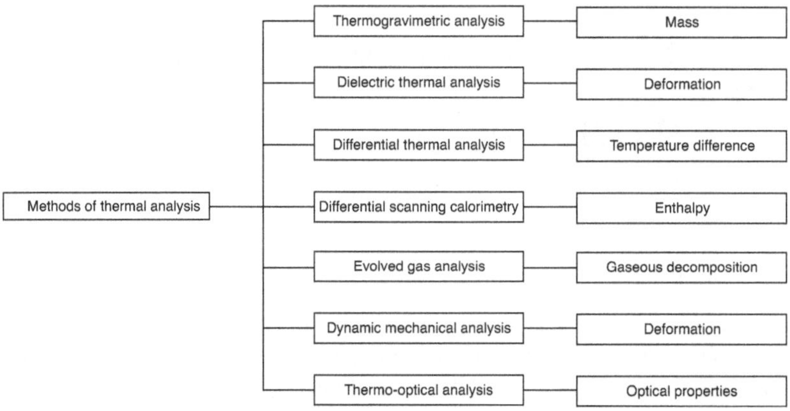

FIGURE 4.1 Types of thermal analysis and associated parameters.

Dielectric thermal analysis is a similar technique in which an oscillating electric field is used instead of a mechanical force to study the physical properties of polar materials. Thermo-optical analysis is a technique based on the optical behaviour of carbon and is used to determine the concentration of elemental carbon in a sample.

4.3 THERMOGRAVIMETRIC ANALYSIS

The uniqueness of the response of a material to variation in temperature makes TGA a widely used technique for the characterisation of materials. Each material has a characteristic temperature at which it is subjected to either physical or chemical changes that result in a weight change. On heating a sample, it will undergo either weight loss or weight gain, depending on the type of thermal reaction. For instance, decomposition, dehydration, and reduction cause weight loss, whereas oxidation and absorption result in weight gain in the sample material. The TGA measurements provide information on physical phenomena such as phase changes, absorption, and crystallisation as well as on chemical phenomena such as chemisorption, oxidation and reduction.

4.3.1 Types of Thermogravimetric Analysis

The experiment is conducted by heating the sample in a controlled environment, and the weight change associated with the temperature rise is recorded. The temperature in the furnace or the heat supplied is the most important parameter. Depending on the amount of heat supplied to the sample, TGA can be classified into three types. They are isothermal thermogravimetry, quasistatic thermogravimetry and dynamic thermogravimetry.

4.3.1.1 Isothermal Thermogravimetry

In isothermal thermogravimetry, the sample is maintained at a constant temperature for the entire duration of weight loss measurement. The change in weight is determined at a uniform temperature, which is decided before the start of the experiment. The weight loss is plotted as a function of time at a constant temperature. Isothermal thermogravimetry is suitable when the study is focused on a particular thermal phenomenon. The oxidative induction time of materials and the amount of moisture content can be determined using isothermal thermogravimetric measurements. The oxidative induction time is defined as the time to initiate the oxidation of a material in an oxidising environment at an elevated temperature. It gives a measure of the thermal stability of the material against oxidation.

4.3.1.2 Quasistatic Thermogravimetry

In this approach, the sample temperature is raised in several steps separated by isothermal intervals, keeping the upper and lower limits for the rate of weight change fixed. The sample is initially heated at a predetermined rate until the upper limit is reached. Thereafter, the sample is maintained at a constant temperature until the lower limit is attained. At this point, the sample would have attained a constant weight. Then, again, the heat is supplied at a linear rate, and the process is repeated.

4.3.1.3 Dynamic Thermogravimetry

This is the most commonly adopted technique for TGA. In this technique, the response of a material when heated at a constant rate is recorded. The temperature increases linearly with time as the experiment progresses. Temperature-dependent phenomena such as evaporation and decomposition can be studied by subjecting the sample to a linear heating rate.

4.3.2 Thermogram

A thermogram or a thermogravimetric (TG) curve is a two-dimensional plot in which weight change is plotted as a function of temperature or time. The weight change is expressed in grams or percentage of grams with respect to the initial sample weight. The shape of a thermogram is characteristic of a thermal reaction and, therefore, can be used for the identification of the compounds present in the sample. Generally, thermograms are categorised into six types based on their shape, as presented in Figure 4.2.

Type a: When there is no weight change for the material in the temperature range chosen for analysis, a flat thermogram is obtained. This is due to the possibility that the thermal stability of the material lies outside the experiment temperature range.

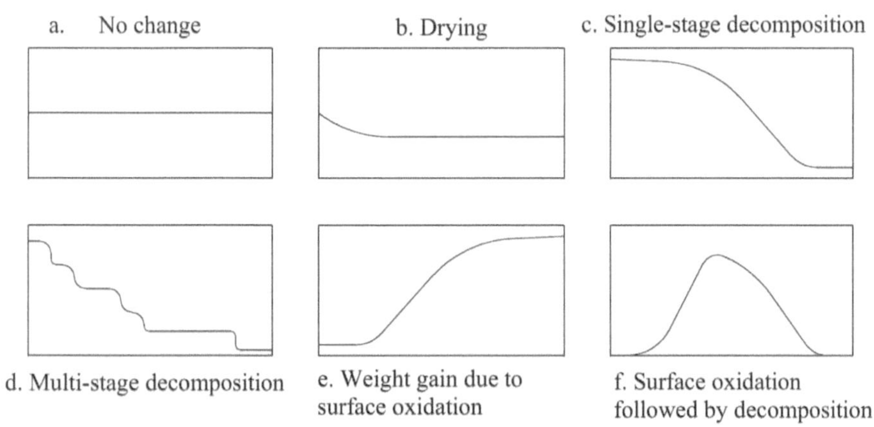

FIGURE 4.2 Different types of thermograms.

Type b: When a weight loss region is followed by a flat line. The region of the flat line is called a plateau. The evaporation of volatile compounds from the sample results in the drying (desorption) of the sample, and this phenomenon is depicted by such a curve.

Type c: Single-stage decomposition or weight loss is represented as a type 3 thermogram.

Type d: Multi-stage decomposition is depicted as a type 4 thermogram.

Type e: This type of curve is observed owing to atmospheric reactions such as the oxidation of the sample. As a result of oxidation, there will be a gain in the weight of the sample.

Type f: In this type, multiple thermal reactions are observed. Initially, the rise in the peak is due to an oxidation reaction, and the further decrease in weight corresponds to the decomposition of the oxidised material.

4.3.3 Derivative Thermogravimetric Curve

A differential thermogram or derivative thermogravimetric (DTG) curve represents the first derivative of weight with respect to time as a function of temperature. The peaks in a DTG curve represent different phenomena, and the area under the curve directly gives the weight change of the sample. The downward and upward peaks in the DTG curve represent weight loss and weight gain, respectively. A DTG curve is usually preferred for quantitative and qualitative analysis of materials since it is easier to interpret a DTG curve compared to a TG curve. Taking the derivative helps to distinguish overlapping thermal reactions and provides a better representation of various reactions occurring in a closer temperature range. Figure 4.3 shows typical TG and DTG curves demonstrating a single-stage decomposition of a material.

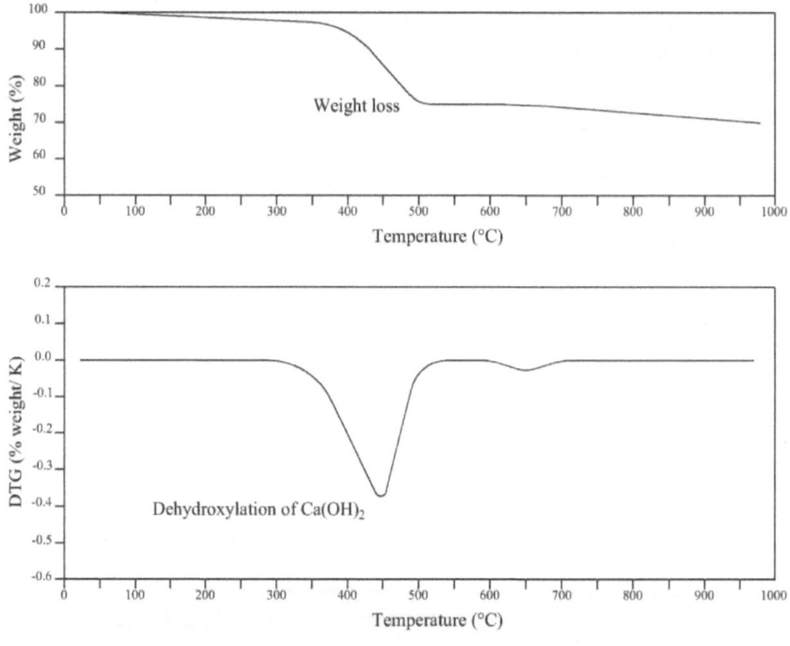

FIGURE 4.3 Typical TG and DTG curve.

4.4 TERMINOLOGIES

Baseline	It is a virtual reference line in a DSC (or DTA) plot that connects two points in the peak region corresponding to the initial and final temperatures as if the enthalpy change is zero (shown in Figure 4.4). A deviation of the curve from the baseline indicates a thermal phenomenon. Depending on whether the peak of the curve is upwards or downwards, the associated reaction can be identified as exothermic or endothermic.
Crystallisation	It is defined as the process by which material gets converted into a crystalline form from a liquid/amorphous state. The crystals formed will have a highly ordered structure.
Crystallisation temperature	It is the temperature at which the nucleation or growth of a crystal initiates. It is a material property.
Endothermic reaction	A reaction that involves the absorption of heat from the surroundings to form products is called an endothermic reaction. Chemical bonds in the material are broken during an endothermic reaction, and for breaking bonds, energy is required. Examples: Melting, evaporation, and sublimation.
Enthalpy	It is defined as the heat content of a system at constant pressure. It is the sum of the internal energy (*U*) of the system and the work done to displace the surrounding environment of pressure "*p*" to make room for the system of volume "*V*". A change in enthalpy is a measure of heat absorbed or released from a system at a constant pressure. It is given by equation 4.1: $$H = U + pV \quad (4.1)$$
Entropy	It gives a measure of the degree of randomness or disorder in a system. An isolated system left for spontaneous evolution will always reach a state of thermodynamic equilibrium where the entropy is highest.

(Continued)

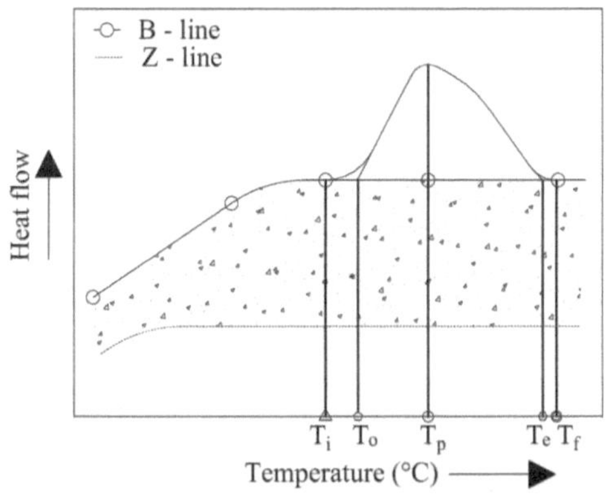

FIGURE 4.4 DSC plot showing baseline, zero line and characteristic temperatures.

Exothermic reaction	A reaction that causes a release of heat along with the formation of reaction products is called an exothermic reaction. The difference in energy between the reactant and the product formed is released in the form of heat. In exothermic reactions, new chemical bonds are formed, which leads to the release of energy. Examples: Crystallisation, freezing, and condensation.
Endset temperature	The temperature at a point where the tangent drawn from the linear descending portion of the peak intersects the baseline is called the extrapolated peak endset (or offset) temperature (T_e).
Gibbs free energy	It is the maximum reversible work that a thermodynamic system can perform in a given initial state at a constant temperature and pressure. It is a measure of the amount of useful work available in a system without increasing the volume or allowing heat to flow from or to the system. The change in Gibbs free energy is given by equation 4.2: $$\Delta G = \Delta H - T\Delta S \quad (4.2)$$ where ΔS is the change in entropy, ΔH is the change in enthalpy, and T is the temperature.
Glass transition	It is a gradual and reversible conversion of amorphous materials from a brittle state into a viscous or rubbery state as the temperature is raised. Heating an amorphous material above the glass transition temperature results in the softening of the material.
Heat capacity	Heat capacity is the amount of heat required to cause a unit change in the temperature of a material. It is a thermodynamic property of a material and is a measure of molecular motion in the material.
Heat flow	Heat is a form of energy that always flows from a body with a higher temperature to a body with a lower temperature. The exchange of energy between two bodies with different temperatures is called heat flow. The flow of energy continues until a state of thermal equilibrium is reached. There are three types of heat flow: conduction, convection, and radiation.
Initial and final temperatures	The temperature at the onset of a reaction is referred to as the initial temperature (T_i) and the final temperature (T_f) marks the completion of the reaction. The DTG curve starts to deviate from the baseline at T_i, increases linearly, reaches a peak maximum, and then descends till it again reaches the baseline at T_f.
Onset temperature	The temperature at a point where the tangent drawn from the linear ascending portion of the peak intersects the baseline is called the extrapolated peak onset temperature (T_o). Unlike the peak temperature, the onset temperature is less dependent on the heating rate. Hence, T_o is more significant in cases where the influence of different heating rates on sample properties is studied.
Peak	A peak in the DSC (or DTA) curve corresponds to a disturbance in the steady state of a material due to an exothermic or endothermic reaction.
Peak maximum temperature	The temperature at which the maximum deviation of the heat flow (or temperature change) curve from the baseline is recorded in a DSC (or DTA) plot is called the peak maximum temperature (T_p). It is the temperature corresponding to the maximum point of an exothermic or endothermic peak.

(*Continued*)

Phase transition	When a thermodynamic system changes from one phase to another due to a variation in temperature and/or pressure, it is called a phase transition. Melting, evaporation, sublimation, condensation, crystallisation, etc. are a few examples of phase transition.
Plateau	It is a region in the thermogram that corresponds to a flat line following a weight loss region. In this portion, there is no reaction, and hence there will be no weight loss or gain for the sample.
Reaction interval	The temperature difference between the starting and ending points of a peak corresponding to a reaction is called a reaction interval. Depending on the number of stages in which a reaction occurs, there can be single or multiple reaction intervals. For example, in single-stage decomposition, there is only one reaction interval, whereas in multi-stage decomposition, the reaction occurs in several steps, resulting in multiple reaction intervals.
Thermal diffusivity	It is a measure of the ability of a material to conduct heat energy relative to its ability to store heat energy. It indicates how fast the heat is transferred through the material.
Thermal lag	It is the difference between the average sample temperature and the temperature measured by the sensor. The sample size and heating rate influence the thermal lag in the measurement. For smaller samples, the thermal lag will be lower since the resistance to heat flow through the sample will be lower.
Zero line	For a well-calibrated instrument, it is the DSC (or DTA) curve corresponding to an empty instrument without samples and crucibles or with empty crucibles.

4.5 INFORMATION FROM THE THERMOGRAM

The thermogram can be useful to acquire the following information:

i. *Thermal stability and kinetics:*

The thermal stability of a material is its ability to maintain its characteristics without much change as the temperature rises. The decomposition of a material in an inert atmosphere is generally studied for predicting the thermal stability of the material. For this purpose, it is essential to optimise the factors that influence weight change during the entire experiment. A good understanding of thermal stability is important to identify the temperature range in which the material can be used. In addition, the kinetics of the various chemical reactions can be investigated using TGA.

ii. *Composition of the material:*

Various compounds present in a material undergo phase changes or other thermal reactions in different temperature ranges. Evaporation, sublimation, decomposition, oxidation, reduction, absorption, desorption, etc. are some of the commonly observed phenomena associated with temperature change. Depending on the type of reaction and the temperature range in which the reaction occurs, the composition of the sample material can be predicted. The compounds can also be quantified by measuring the weight loss/gain of the sample with the temperature rise.

iii. *Oxidative stability:*

Oxidative decomposition of organic substances and oxidation of metals in air or oxygen can be studied using TGA. Instead of inert gas, air or oxygen is made to flow through the furnace to create an oxidative atmosphere.

iv. *Moisture and volatile content of materials:*

Quantification of bound water in cement paste is mainly done using TGA. Besides, TGA can be applied for the analysis of gaseous or volatile products that evolve during the reaction of the material. It has to be coupled with other instruments, such as a mass spectrometer, for the evaluation of volatile compounds.

v. *Purity of the material:*

The purity of materials can be checked by performing TGA on samples. Additional or unusual peaks in the curve that are not characteristic of the material are indicative of the presence of impurities in the sample. Other characterisation techniques can be employed to identify the impurity.

4.6 INSTRUMENTATION

A thermogravimetric analyser is an instrument used for conducting TGA. A typical TG analyser consists of a sensitive microbalance, a furnace in which the crucibles are placed, a temperature controller that controls the temperature in the furnace, and a purge gas for creating an inert atmosphere, as presented in Figure 4.5.

4.6.1 Analytical Balance

A highly precise and accurate microbalance is used to measure the change in weight of the sample. The balance is designed in such a way that an electrical signal proportional to the weight change of the sample is generated. The electrical signal is converted into weight loss with the help of a data acquisition system. An ideal balance should record the weight change of a sample accurately and have an immediate response to the weight change. It is important to note that the balance must always be thermally isolated from the furnace. The balance should be unaffected by vibrations and temperature changes.

There are two types of balances: deflection-type and null-type balances.

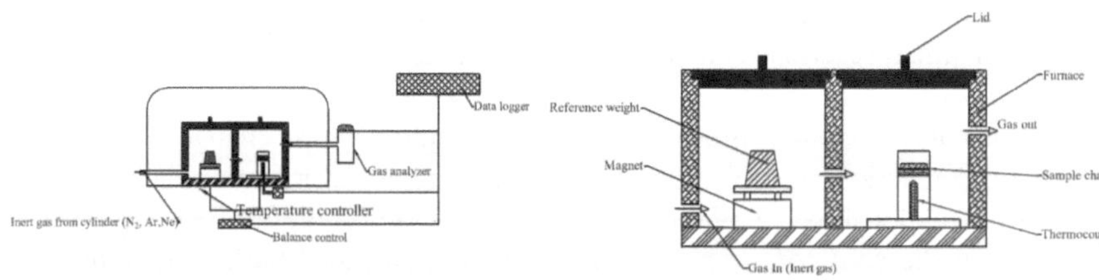

FIGURE 4.5 General arrangement of components in a TG analyser.

4.6.1.1 Deflection-Type Balance

Deflection-type balances are those in which deflection or displacement of the instrument is the basis for measurement of the weight change (Figure 4.6). The deflection-type balances are further classified into four different types:

i. *Beam type:*

The weight change of the sample causes the balance beam to deflect, which is then recorded using various methods and used to calculate the weight change. One of the methods is to use an optical lever arrangement. In this arrangement, a mirror is mounted on the beam, and the intensity of light reflected from the mirror is recorded. The variation in the intensity of light is proportional to the deflection of the beam, and it gives a measure of weight change. Another method to measure the beam deflection is to process the signals generated by a displacement transducer attached to the beam. The most commonly used transducer for converting the linear motion of the beam into a corresponding electrical signal is a linear variable differential transformer (LVDT). Strain gauges are also used to measure the deflection of the beam.

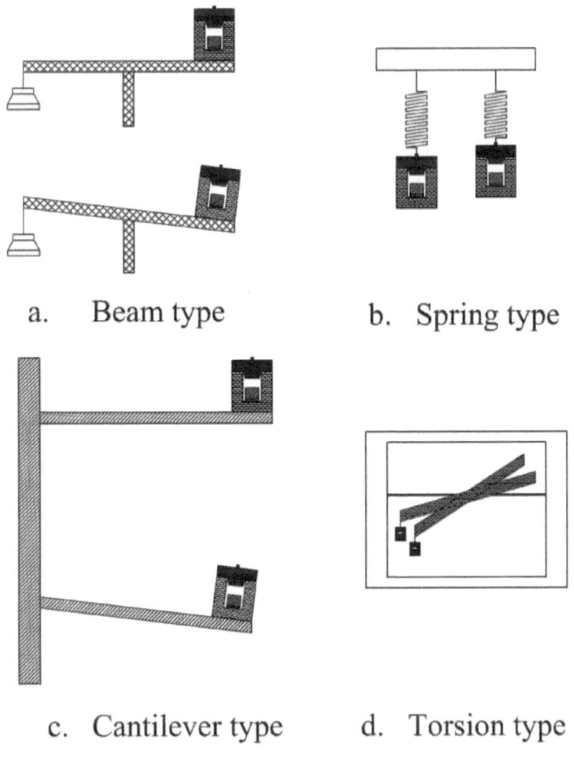

FIGURE 4.6 Deflection-type balances.

ii. *Helical spring type:*

This balance works on the principle of Hooke's law. The elongation or contraction of the spring in the balance is recorded to determine the weight change of the sample. The material of the spring has to be chosen carefully; it should be inert to temperature changes and shouldn't have problems related to fatigue.

iii. *Cantilever type:*

A cantilever-type balance has one end fixed and the other end free. The sample is placed at the free end so that it is free to displace. The deflections of the cantilever are recorded and used to calculate the weight change.

iv. *Torsion type:*

The balance beam on which the specimen is loaded is attached to a taut wire, which acts as a fulcrum. The wire will be firmly attached at one end or both ends. According to the weight change, the beam will deflect, and the deflection is proportional to the torsional characteristics of the wire.

4.6.1.2 Null-Type Balance

The null-type balance is the most preferred type, as it is more accurate and highly sensitive (Figure 4.7). It mainly consists of a beam, a light source, an aperture, two photodiodes, a coil, and two permanent magnets above and below the coil. Initially, the beam is in a null position, and the light passing through the aperture attached to the beam strikes the two photodiodes in equal amounts. As the experiment proceeds, the sample loses weight and becomes lighter. The side of the beam holding the crucible will deflect upwards, and correspondingly, the reference weight will move downward. When the beam is out of position, the light rays deviate from their original path, and unequal amounts of light reflect on the photodiodes. The photodiodes convert the incident light energy into electrical energy (current). An electric current proportional to the deflection of the beam is passed through the coil (electromagnet). The electromagnet acts as a pivot for the balancing beam, and the electric current interacts with the magnetic field to produce a restoring force that pushes the reference mass upwards and restores the beam to its null position. The force required to restore the beam to its original position depends on the amount of current passed through the coil, which in turn gives a measure of the weight loss of the sample.

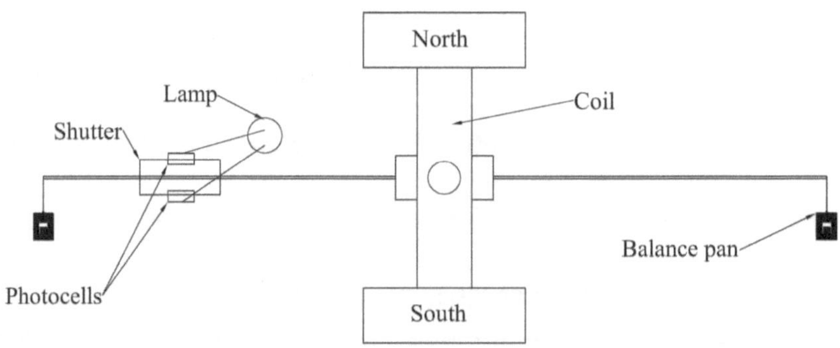

FIGURE 4.7 Null-type balance.

4.6.2 Crucible

Crucibles are small containers in which the samples under study are placed. The crucible is attached to the weighing arm of the balance either by fixing it or by hanging it down using a suspension wire. One of the requirements of the crucible is to transfer the heat uniformly to the sample. Crucibles made of materials with high thermal diffusivity are preferred because they conduct heat rapidly. Care should be taken that the material of the pan is not reactive to the sample material. Hence, a pan with suitable material has to be chosen depending on the temperature range, nature, and amount of the sample. Some of the crucibles can influence the thermogram by acting as catalysts for the thermal reaction of the sample material.

The commonly available crucibles are made up of aluminium, platinum, alumina, copper, stainless steel, gold, glass, ceramics, etc. The material used for the crucible is selected based on the temperature range in which the experiment is performed. The crucible must be thermally stable at a temperature at least 100°C higher than the temperature condition set for the experiment. Disposable pans made of aluminium are available for experiments that involve the generation of residues that stick to the pan and are difficult to remove. However, it should be ensured that the maximum temperature of the experiment is below the melting point of aluminium, which is 600°C. In general, powdered samples are analysed using flat crucibles and liquid samples using deep crucibles. Depending on the intended purpose, crucibles are of the following types (shown in Figures 4.8 and 4.9):

i. *Shallow-type crucible:* This type is used when diffusion of volatile compounds or gaseous products formed during the testing is a rate-controlling factor. These volatile products should be able to escape immediately after they are formed so that they can be recorded as weight loss. In such cases, the sample used for testing is prepared in the form of a thin layer so that the gaseous products can easily escape from the shallow pan.

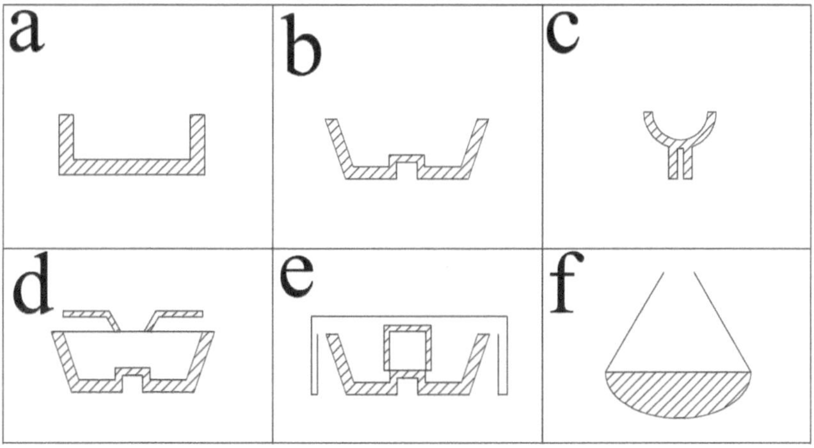

FIGURE 4.8 Different shapes of crucibles.

FIGURE 4.9 Different shapes of crucibles.

ii. *Deep crucible:* It is used for studies where side reactions are important or when fluid samples are studied.
iii. *Loosely covered crucible:* It is preferred for studies involving isothermal heating of the sample. In such cases, the rate of heating and weight loss of the sample are of less significance.
iv. *Flat-plate type crucible:* These are used for high vacuum studies or in cases where a large surface area of exposure is required.
v. *Polyplate crucible:* For large quantities of samples and an extremely large surface area, this type of crucible is used.

4.6.3 Furnace

In a typical TG instrument, the sample pan and the microbalance are placed inside a furnace. The furnaces are mostly electrical resistance heaters. In some TG instruments, infrared or microwave heating is also used to obtain highly rapid heating and cooling rates. The heating coils in the furnace are non-inductively wound so that there is no magnetic interaction between the coil and the samples inside the furnace. The presence of magnetic interactions can affect the measurement of the weight of the sample. The furnaces are available in a specific range of temperatures, from −150°C to 2,800°C. In general, there are four different furnaces designed for a particular temperature range within this limit: −150°C to 500°C, 25°C–1,200°C, 25°C–1,600°C, and 400°C–2,800°C. The most commonly adopted furnace is the second type. The higher the temperature range of the furnace, the higher the cost of the instrument. Moreover, the temperature range of the experiment governs the type of material to be chosen for the heating coil. Up to a temperature of 1,750°C, platinum alloys are commonly used, whereas for temperatures above 1,750°C, special furnaces made of tungsten, molybdenum, or graphite are preferred.

Some of the important features of a furnace are as follows:

i. Furnaces have a hot zone of uniform temperature for a sufficient length where the crucible and temperature sensors are located. For a deflection-type balance, the zone must be large enough to accommodate the movement of the beam.

ii. The selection of the furnace should be such that it suits the temperature regime to be used for the experiment.
iii. The heat generated from the furnace should not affect the functioning of the balance.
iv. It should be so designed that the required heating rate can be easily achieved and a rapid response to heating is possible.
v. The maximum working temperature of the furnace must be at least 150°C higher than the desired temperature range for conducting the experiment.

The position of a furnace with respect to the balance has a significant influence on the results obtained. The balance and furnace have two possible configurations: vertical (Figure 4.10a and b) and horizontal (Figure 4.10c). In the vertical configuration, there are two positions of the sample in the furnace: suspended or hanging position (a) and top loading (b) as shown in Figure 4.10. In top loading, the crucible is supported above the balance by a vertical rod or stem, and in bottom loading, the crucible is hung down below the balance. Top loading is preferred for higher temperature ranges, whereas for lower temperatures, a furnace below the balance is preferred. In the horizontal configuration, the crucible is attached very rigidly to the end of a balance arm, and it is a side-loading arrangement.

The limitation of vertical configuration is the buoyancy effect due to variation in the density of gas with temperature rise. The upward force (buoyancy) exerted by the surrounding atmosphere on the sample results in an apparent gain in weight. Hence, a correction for the buoyancy effect should be applied, without which the weight recorded will be higher than the actual weight of the sample. For this purpose, an empty crucible used for the experiment is initially kept in the furnace under the same conditions as the actual experiment and subjected to a temperature rise. The measurement is recorded, and correspondingly, a blank curve is obtained, which is then subtracted from the actual experimental curve.

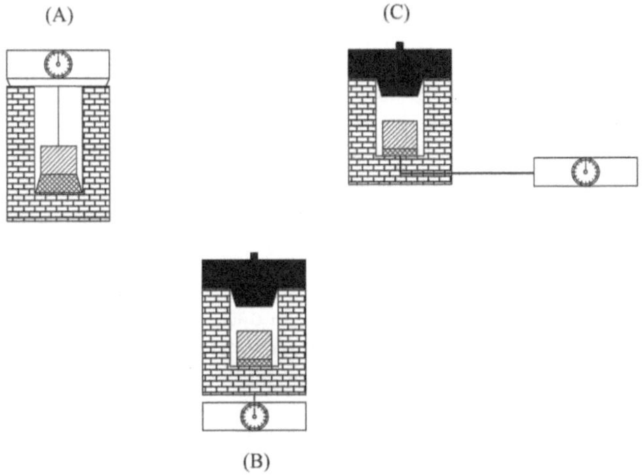

FIGURE 4.10 Various configurations of balance and furnace.

4.6.4 Temperature Programmer

The temperature of the furnace is measured using temperature sensors such as thermocouples. The temperature recorded by the sensor is transmitted to the programmer, and the programmer compares that temperature with the value set by the operator. Depending on the temperature requirement, the power supply to the furnace is monitored. For instance, if the temperature in the furnace is lower than that set by the operator, the controller sends a signal to supply more power to the furnace.

A thermocouple is an electric device that consists of two dissimilar electrical conductors forming an electric junction at differing temperatures. The temperature difference at the junction generates a thermoelectric current. The temperature-dependent potential developed as a result of the thermoelectric effect is recorded to calculate the temperature. The relation between potential or emf (E) and temperature (T) is shown in equation 4.3, where S_1 and S_2 are known as Seebeck coefficients:

$$E = \int_{T_1}^{T_2} (S_1 - S_2)\, dT \tag{4.3}$$

Fabrication of thermocouples is done using various materials such as tungsten, aluminium, and platinum alloys. Aluminium alloys such as chromel and alumel are used for temperatures below 1,100°C, platinum alloys are used up to 1,750°C and tungsten is used for higher temperatures. The position of the thermocouple should be as close to the sample as possible. There are different positions at which a thermocouple can be placed near a crucible depending on the configuration of the furnace and balance, as presented in Figure 4.11. In side and hanging configurations of the balance, the sensor is placed close to the crucible and not directly in contact with it to avoid interference with weight measurements. A thermal lag in the measurement of temperature is observed due to a lack of direct contact. However, in the top-loading arrangement, the sensor is placed in direct contact with the crucible.

4.6.5 Recorder

The output from a microbalance is recorded using a recorder. Some of the commonly used recorders are chart recorders, microcomputers, X–Y recorders, and time-base potentiometric strip chart recorders. On a microcomputer, the thermogram is plotted using preinstalled software. Most of the recorders record a change in the weight of the sample, whereas others directly give the percentage weight change. To calculate the percentage weight change from the change in weight data, the initial weight of the sample must be known. Data acquisition systems are generally run along with the recorder to have visual monitoring of the experiment and analyse the data. Modern data acquisition systems have many additional features, such as curve smoothing, differentiation, and integration of data.

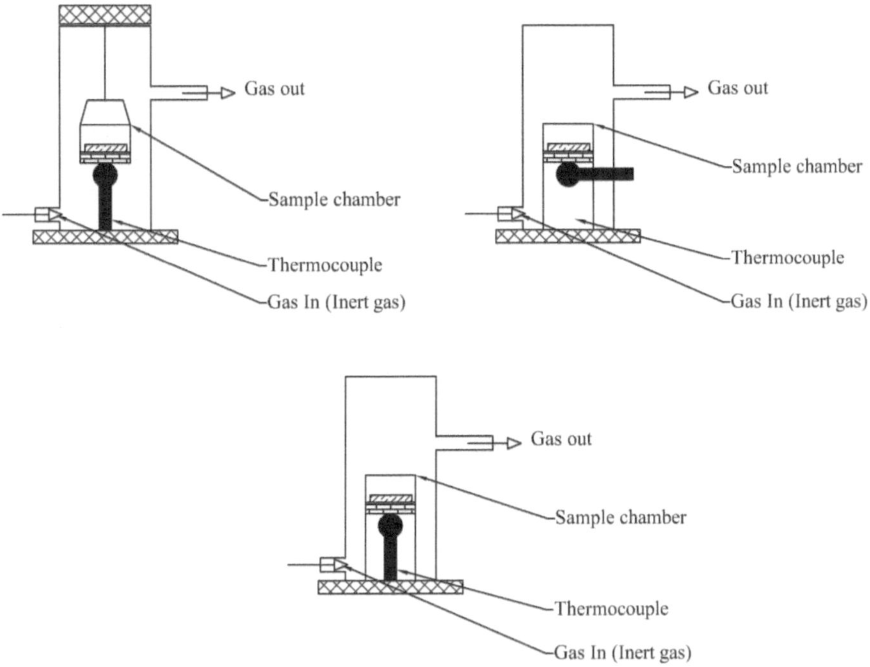

FIGURE 4.11 Typical locations of a thermocouple.

4.6.6 Purge Gas

The purge gas is made to flow through the furnace to remove the gaseous reaction products that evolve during the experiment and prevent the deposition of the reaction products inside the furnace. It is used as a protective gas to protect the balance against the action of corrosive gaseous products that are evolved. These gases also help with faster heat transfer to the sample. Inert gases such as nitrogen, argon, or helium are commonly used as purge gases when it is essential to create a non-reactive atmosphere. It will ensure that the sample will respond only to changes in temperature. The use of helium as a purge gas is known to conduct more efficient heat transfer from the furnace to the sample, especially at temperatures below 700°C. Moreover, purge gases act as a smooth thermal blanket, which helps to create a uniform temperature region and avoid localised hotspots. They also help to cool down the furnace so that faster cooling rates can be achieved. In some cases, reactive gases such as air, oxygen, or hydrogen are also passed through the chamber to aid the desired reaction or act as catalysts. Air, or oxygen, is oxidising in nature, whereas 8%–10% hydrogen mixed with inert gas is used to create a reducing atmosphere. To prevent the possibility of an explosion, high concentrations of hydrogen are not used, and if required, special care needs to be taken.

4.7 WORKING PRINCIPLE

The basic functioning of a TG instrument is described in this section. The sample kept inside the furnace is heated by a resistance heater, infrared radiation or microwave radiation. The temperature of the sample is recorded by a temperature sensor, commonly, a thermocouple. The recorded temperature data is sent to a data acquisition system (computer) for processing. The computer then sends the information to the temperature controller, and the controller passes a signal to raise the temperature in the furnace at a controlled rate. During the temperature rise, the sample will undergo weight loss/gain due to various physical and chemical changes that occur to the material. Thermal reactions such as decomposition, dehydration, dehydroxylation, reduction, and desorption result in weight loss, whereas oxidation and absorption result in a gain in the weight of the sample. When a null-type balance is used to measure the weight change, a restoring force (either mechanical or electrical) is applied to the balance beam that is equal to and opposite to the weight change. Thereby causing the balance to return to its null position. The weight change is plotted on the *y*-axis, and the temperature recorded is plotted on the *x*-axis to obtain the thermogram. The weight change may be expressed in either milligrams or percent of the original specimen weight.

4.8 INFLUENCING FACTORS

The TG measurements obtained have a great deal of dependence on various specimen characteristics and instrument factors. Furnace atmosphere, heating rate, particle size of the sample, sample size, gas flow rate, etc. are some of the factors that influence the results. Due to this reason, measurements made at one laboratory cannot be compared with those obtained at another laboratory. At least one of these will vary from one setup to another. Hence, it is required to always stick to the same procedure and experimental conditions and to carry out the experiment on the same day for all the samples under study.

4.8.1 SPECIMEN CHARACTERISTICS

4.8.1.1 Particle Size of Sample

For certain reactions, such as those that involve the diffusion or evolution of gaseous products, the size and shape of the sample are quite significant. Samples with smaller particle sizes have a larger surface area and aid in the easy diffusion or evaporation of gases. The rate of weight loss depends on both the formation of gaseous products and the ease of diffusion of gases through the sample. Coarser particles result in a slower rate of reaction and delayed peaks in the DTG curve.

4.8.1.2 Packing Density

The packing density also has an impact on the weight loss rate of a material. A sample that is densely packed will have higher thermal conductivity and, hence, result in faster heating of the sample. On the other hand, loose particle packing in the sample results in higher permeability. This leads to better diffusion of gaseous reaction products and a faster rate of weight loss.

4.8.1.3 Sample Size

The quantity of sample (sample size) used for the experiment influences the rate of weight change as it affects the rate of heat transfer and the diffusion of gaseous products into the atmosphere. In most cases, smaller sample sizes result in faster heating of the sample since it reaches equilibrium with the furnace temperature quickly. For non-homogeneous materials and materials with low volatile content, a sufficiently large quantity of samples must be taken to obtain good results. This is because the signals produced from small samples (<1 μg) will be weak, and for non-homogeneous materials, each sample part may react differently. A larger sample size also leads to a broader peak in the DTG curve and a shift in the peak to a higher temperature. For example, dehydration of gypsum occurs at 115°C for a small sample size, but the peak shifts to 145°C if the quantity of gypsum is higher. This is because the formation of higher vapour pressure due to the evaporation of more water from a large quantity of gypsum requires a higher temperature for the complete dehydration of gypsum. However, too large a sample size is also not preferred, as it results in a higher thermal lag and temperature gradient in the sample. The temperature gradient is defined as a non-uniform distribution of temperature within the sample, which depends on the sample size, heating rate, and thermal diffusivity of the sample.

4.8.2 Instrument Factors

4.8.2.1 Furnace Atmosphere

For measuring the weight change as a function of temperature, the sample should be free to exchange gaseous products that are formed with its immediate surroundings. The transfer of heat to the sample and the thermal reactions of the material depend on the furnace atmosphere to a certain extent. Some of the materials may remain stable in air, whereas other materials, such as metal, will undergo oxidation in the presence of oxygen. In the case of cement, carbonation is also a problem when the atmosphere surrounding the sample is air. Hence, the kind of atmosphere inside the furnace has to be chosen so that it suits the requirements of the experiment. Air, nitrogen, and argon have lower thermal conductivity, whereas helium has high conductivity. The higher the thermal conductivity of the atmosphere, the faster the heat transfer. For instance, helium has a higher thermal conductivity than carbon dioxide and therefore transfers heat by conduction nearly eight times faster. Besides, at times it is required to conduct the experiment at reduced pressure or by creating a vacuum in the furnace. This is to separate the overlapping peaks of decomposition and evaporation of a material.

4.8.2.2 Furnace Size

The size of the furnace is another factor that influences the thermogravimetric measurements. In small furnaces, the desired heating can be achieved in a short time. However, it is very difficult to control the rate of temperature rise, which result in non-uniform heating of the sample. On the other hand, large furnaces will require more time to reach the targeted temperature. Furnaces with larger sizes will ensure proper maintenance of a uniform hot zone and will be able to withstand higher temperature regimes.

4.8.2.3 Gas Flow Rate

The TG experiments are always performed in a flowing gas stream, and the flow rate of the gas affects the rate of heat transfer. 30 ml/min is the typical flow rate adopted for purge gases in TG experiments. A static atmosphere or a very low flow rate will not help with heat transfer or removal of gaseous products. However, very high flow rates of gases are also not advisable since they can disturb the balance mechanism. Besides, at higher flow rates, the temperature at which weight change occurs is lowered, and the different peaks in the thermogram will be difficult to distinguish.

4.8.2.4 Crucible Properties

The material of the crucible must be chosen such that it will not react with the sample and affect the thermogravimetric measurements. The thermal conductivity of the crucible material influences the heat transfer from the crucible to the sample. The higher the thermal conductivity, the greater the rate of heat transfer. Alumina pans are typically used for thermal analysis, whereas platinum pans are preferred for temperatures above 640°C. Owing to its higher thermal conductivity than alumina, TGA measurements obtained using platinum crucibles are better. Moreover, for the reactions that require platinum as a catalyst, pans made of platinum are used. Gold crucibles are chemically resistant; however, their use is limited due to their high cost. Copper crucibles are exclusively used for determining oxidative stability, for which copper acts as a catalyst. Sapphire crucibles are highly resistant and suitable for the analysis of materials with high melting points.

The geometric shape of the crucible is also significant as it regulates the rate of diffusion of volatile reaction products into the surroundings. A shallow crucible will allow faster diffusion of the gases, whereas a narrow and deep crucible will restrict the easy escape of the products.

The TGA measurements also depend on whether the crucibles are kept open to the atmosphere or sealed airtight with a small opening in the lid. If sealed crucibles are used, then the weight change is observed to shift to a higher temperature. Moreover, the DTG peaks corresponding to different reactions that take place at a close range of temperatures are observed to be well separated. When the lid is kept closed, the gaseous products formed can escape only through the small hole, which leads to higher vapour pressure and a slower rate of reaction. On the other hand, certain reactions such as oxidation, reduction, and dehydration require a good interaction with the surroundings. In such cases, the maximum exposure to the sample can be provided by placing the sample on a thin bed with the purge gas passing through it. For example, dehydration of gypsum generally occurs in two stages, and there are two peaks observed in DTG corresponding to dehydration of gypsum to hemihydrate and hemihydrate to anhydrite. In the case of an open vessel, both peaks appear at around 140°C with not much separation. On the contrary, for a closed vessel, two distinct peaks and a shift in the peaks to higher temperatures can be clearly observed, as shown in Figure 4.12.

4.8.2.5 Heating Rate

The temperature at which a material reacts and undergoes a physical or chemical change is dependent on the heating rate of the sample. The typical rate at which heat

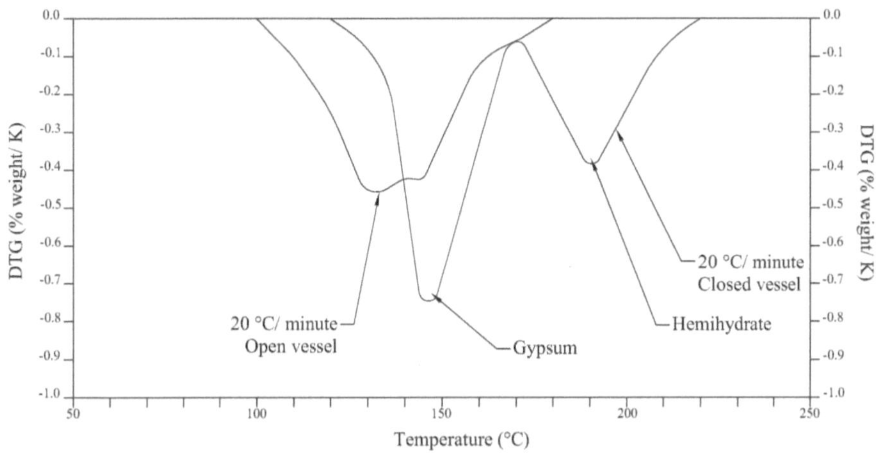

FIGURE 4.12 Influence of open and closed crucibles on thermogram.

is supplied lies in the range of 5°C/min–20°C/min. The heating rate should be chosen carefully; it should neither be too slow nor too fast. Even though slower heating of the sample results in better separation of the overlapping peaks, too slow a rate will consume more time for the completion of the experiment. At faster heating rates, the reactions tend to occur at higher temperatures owing to the higher vapour pressure that is developed over the sample. Hence, a shift in the associated peak is observed in the DTG curve. Rapid heating of the sample might also lead to its melting before it undergoes decomposition. Besides, the thermal lag observed during the transfer of heat from the furnace to the sample is greater at higher heating rates. For isothermal thermogravimetry, the sample is heated very rapidly to reach the required static temperature to avoid the occurrence of thermal reactions during the heating.

4.9 EXPERIMENTAL PROCEDURE

The experimental procedure, including specimen preparation, testing conditions and a step-by-step procedure for conducting TG experiments, is explained in this section.

4.9.1 Sample Preparation

The samples used for the TGA should be preferably small, powdered and evenly spread in the crucible. It should be representative of the material under study, and the weight of the sample should be sufficient for accurate measurement. The nature of the sample governs the result obtained from TGA to a great extent. Especially in cases where surface reactions are involved, the behaviour of crystalline samples is very different from that of powdered samples.

4.9.1.1 Pretreatment of Sample

Pretreatment methods for hydrated cement pastes are described in this section. For the thermal analysis of hydrated cement pastes, pretreatment is one of the most significant steps to getting the desired results. This is done to ensure that all the free water present in the pores of cement paste is removed and the hydration of cement can be arrested. For instance, if we want to study the properties of hydrated cement at a specific state, say after 28 days of hydration, then the hydration has to be stopped at 28 days. For this purpose, the sample has to be treated prior to testing using solvent exchange or freeze-drying methods. It should be ensured that the sample is analysed immediately or on the same day so that the sample is unaffected by the surrounding conditions. Otherwise, the sample has to be carefully stored in an airtight container under an inert atmosphere or a light vacuum until it is taken for the experiment.

i. *Solvent exchange method:*

The solvent exchange method is one of the most widely adopted methods for arresting cement hydration. Alcohols such as methanol, ethanol, isopropanol and acetone are commonly used as solvents for the removal of evaporable water. These solvents should be miscible with water but have a higher vapour pressure. The steps involved in the solvent exchange method are mentioned below:

Step 1: Powdered sample is prepared by crushing and grinding it using a mortar and pestle. The grinding of the sample is carried out in a nitrogen atmosphere to minimise carbonation. After grinding, the sample is made to pass through a 75 μm sieve.

Step 2: Approximately 5 g of the powdered cement paste is immersed in a glass beaker with 50 ml of the solvent.

Step 3: The suspension is rested for 10–15 minutes for the exchange of the free water present in the sample with the solvent. Isopropanol is commonly used as the solvent compared to other alcohols such as methanol and ethanol due to the lower chance of it getting absorbed by the cement hydrates.

Step 4: After 10–15 minutes, the suspension is filtered to remove excess alcohol. For further removal, 5–10 ml of diethylene ether is passed through the sample remaining after filtration.

Step 5: The next step is to dry the sample for 8–10 minutes in an aerated oven at 40°C or in a vacuum desiccator. Another option is to pass nitrogen through the solid until it dries and attains a constant weight.

Limitations of this method:

- In addition to the removal of free water, the organic solvents may replace some quantities of bound water.
- Storage of the cement paste sample in methanol, ethanol, or isopropanol for a prolonged duration affected the structure of ettringite and monosulphoaluminate. It caused the dehydration of monosulphoaluminate from 12 to 10 molecules of water. Moreover, immersion in methanol led to a complete breakdown of the ettringite structure, whereas the use of ethanol or isopropanol as solvent resulted in little loss of water.

- Among the organic solvents used, methanol has the greatest ill-effect on the TG data. Methanol is reported to interact with CH and C–S–H to form calcium methoxide. At temperatures above 200°C, the thermal decomposition of methanol causes the release of CO_2, which then results in the carbonation of portlandite. Hence, owing to the sorption of methanol, there is a reduction in the CH peak and an increase in the weight loss corresponding to carbonates.

Solution:

To avoid the sorption of these alcohols by the hydrated products, a less polar solvent such as diethyl ether can be added to remove the alcohols after the exchange of free water from the paste.

ii. *Freeze-drying method:*

Quenching the cement paste sample in liquid nitrogen followed by freeze-drying is a preferred pretreatment method in cases where carbonation of the sample should be prevented. In this method, the sample is first immersed in liquid nitrogen for around 15 minutes and then dried in a freeze dryer. Vacuum drying is adopted to obtain a CO_2-free environment where the frozen water is directly converted into water vapour. The dried sample is then crushed into powder before the experiment.

Limitations:

- Excessive and prolonged drying will cause dehydration of ettringite. Crystalline ettringite will generally have 30–32 moles of water, and on drying, it is reduced to another form called meta-ettringite, which has only 9–13 moles of water.
- C–S–H and monosulphoaluminate are some of the other hydrates that are affected by drying for prolonged durations. Some of the chemically bound water is removed from C–S–H and monosulphoaluminate on excessive drying.

Solution:

Prolonged durations of drying will affect the hydration products, and short-term drying alone is not sufficient to remove all the free water. Hence, a two-step procedure involving drying for a shorter duration combined with solvent exchange can be adopted for arresting the hydration of cement paste as follows. The sample is immersed in alcohol to remove free water, followed by passing diethyl ether through the sample to replace alcohol, which may stay absorbed into the hydrated phases. To remove the diethyl ether remaining in the pores of the sample, dry the sample at 40°C after the solvent treatment.

4.9.1.2 Points to Note

i. The possibility of carbonation in hydrated cement is an important factor to be considered during the preparation stage of the sample. Exposure to air during the preparation or storage of the sample will cause CO_2 in the atmosphere to react with the hydration products such as $Ca(OH)_2$ and C–S–H present in the hydrated cement. The large surface area of powdered samples makes them more susceptible to carbonation. Moreover, the maximum rate of carbonation is observed at a relative humidity of 50%–60% which is present during the drying stage. Hence, utmost care should be taken to avoid contact of the sample with CO_2 during drying. To reduce carbonation, the powdered and dried samples are stored in airtight containers with nitrogen or a vacuum inside.

ii. The drying method adopted depends on the hydration products that are of interest for the analysis. For example, drying for a short duration is preferred for the study of ettringite and monosulphoaluminate since extensive drying will result in the dehydration of these phases. On the other hand, for the quantification of bound water, drying the sample in a dryer at 105°C is not preferred, and instead, the solvent exchange method is adopted. This is because ettringite, C–S–H, and monosulphoaluminate start losing bound water around that temperature.

4.9.2 Testing Conditions

Crucible type, the environment in the furnace (inert or oxidative), temperature programme, sample size, etc. are a few parameters to be decided by the operator before performing a TG experiment. Setting these parameters along with proper calibration is required to ensure good reproducibility of test results. Good reproducibility is important to compare different datasets of the same material. A few important factors to be considered for TGA are mentioned below:

i. The quantity of the sample should be between 2 and 50 mg. To ensure reproducibility, the sample size should be maintained the same for all samples during each experiment. The sample should be spread uniformly in the crucible to ensure an even distribution of temperature throughout the sample.

ii. A crucible made up of a suitable material and geometry has to be chosen for placing the sample. Alumina or platinum crucibles are the most preferred. For 50 mg of sample quantity, a crucible of 150 μl volume is typically used.

iii. The heating rate of the sample is chosen such that the duration of the experiment is within a certain limit and overlap between reactions in the thermogram is avoided. The temperature ranges in which the reactions of our interest occur should also be taken into consideration. The typical heating rate adopted is 20°C/min for heating a sample to 1,000°C.

iv. The purge gas adopted should be suitable for the thermal reactions that may occur during the experiment. Generally, it should be inert to ensure that the weight change of the sample is due to temperature change alone. For cement paste, nitrogen is commonly used as a purge gas. The gas flow rate should also be selected carefully, and the typical range is 30–50 ml/min.

4.9.3 Stepwise Procedure

The basic steps to be followed for the TGA of hydrated cement paste are given below:

Step 1: Ensure that the walls of the furnace and the crucible are clean and free from residues from the previous run before the start of the next run. The furnace and crucibles are to be periodically heated to 800°C in the air for 10–15 minutes for cleaning. Solvents can be used to remove condensed reaction products that are stuck on the furnace walls. If the products are too stubborn, then bake the furnace in hot air at a temperature above 600°C and then scrub the inner surfaces with a fine-grit emery sheet.

Step 2: Place the empty crucible in a stirrup using forceps or crucible tongs. The stirrup with the crucible is loaded on the autosampler or sample tray. The position of the crucible in the tray should be noted.

Step 3: In the software, make sure that the parameters for the experiment, such as type of purge gas, heating rate and crucible type, are entered correctly. The initial and final temperatures of the experiment are also set in the software.

Step 4: Once the parameters are set, start the experiment. The autosampler will automatically rotate and place the crucible onto a hook hanging down from the microbalance in a bottom-loading arrangement. The rotation of the sample tray is to be done manually in some cases.

Step 5: After the successful pickup of the crucible, the furnace is raised to the desired position. It should be taken care that the sample is freely hanging inside the furnace. After the furnace is in place, the balance will set the weight of the empty crucible to zero.

Step 6: Once the weight of the crucible is tared to zero, initiate the heating programme within the required temperature range. Record the temperature and corresponding weight change of the empty crucible. The blank curve is plotted using the TG data obtained.

Step 7: This process takes about 2 minutes. After that, the furnace is again lowered, and the crucible is automatically unloaded.

Step 8: Take out the stirrups from the sample tray using forceps. Add 5–50 mg of the sample to it carefully. Remember to spread the sample evenly in the crucible. For liquid samples, a small syringe can be used to load the crucible. Check the weight of the crucible with the sample. A variation of ±1 mg from the desired weight is acceptable.

Step 9: Load the crucible with the sample in the same position in the autosampler where the empty crucible was placed earlier. Repeat the same steps that were performed for the empty crucible.

Step 10: The thermogravimetric data is collected and processed by the software, and the thermogram is plotted. The curve obtained is corrected by subtracting the blank curve to compensate for the apparent gain in weight due to the buoyancy effect.

4.10 ADVANTAGES AND LIMITATIONS OF TGA

One of the main advantages of using TGA is the potential of the method to characterise amorphous materials such as C–S–H and aluminium hydroxide that cannot be detected using X-ray based techniques. Hence, the method is most often used in conjunction with other techniques such as XRD and SEM for identifying and quantifying the minerals present in the cementitious system. In cement chemistry, TGA is most commonly used for quantifying bound water, portlandite, and carbonates present in the cementitious system. TGA is sensitive to detecting even small quantities of portlandite and carbonates. However, the method is limited to detecting only those phenomena that are associated with weight change. For instance, the dehydration of C–S–H, which is the main hydration product in cement, happens over a wide temperature range from 50°C to 600°C. This overlaps with the weight loss regions of ettringite, portlandite, and other notable hydrates, making deconvolution of overlapping peaks of hydrated cement difficult. Another limitation is that TGA alone cannot be used for the identification of volatile products that are released from the sample material when heated. It has to be coupled with other techniques such as mass spectroscopy or Fourier transform infrared spectrometry (FTIR) for this purpose.

4.11 OTHER THERMAL ANALYSIS METHODS

Besides TGA, DTA and differential scanning calorimetry (DSC) are the two methods commonly adopted for the thermal analysis of a material. When a material is heated, the enthalpy changes as a result of a physical or chemical change in the material. Some of the phenomena that cause enthalpy change are melting, evaporation, sublimation, crystallisation, etc. DSC measures the difference in heat supply required for the sample to achieve the same temperature as that of the inert reference material used for the analysis. On the other hand, DTA determines the temperature difference between the sample and reference materials, both of which are heated identically. The difference in temperature arises due to phase changes such as melting, evaporation and sublimation, and other thermal reactions that result in enthalpy changes. A representative DTA curve showing various thermal phenomena is presented in Figure 4.13. Exothermic reactions such as crystallisation and oxidation result in a higher temperature of the sample than the reference and hence have a positive peak, whereas melting, which is an endothermic reaction has a negative peak in the DTA curve.

The reference material should be carefully chosen depending on the operating temperature range, thermal conductivity and inertness of the material. Some of the materials used as a reference for DTA and DSC measurements are alumina, carborundum and SiC. In some cases, an empty pan is used as the reference for the experiment. If the operating temperature is below 500°C, aluminium pans are used, and for temperatures above 500°C, gold or graphite pans are suitable. The reference materials should have the following characteristics:

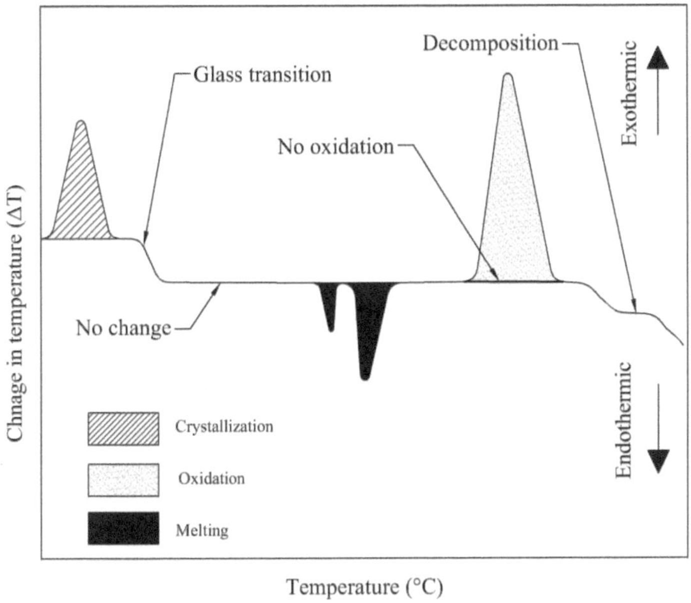

FIGURE 4.13 Representative differential thermal analysis curve for various thermal reactions.

i. The material chosen should be inert, i.e., it shouldn't undergo thermal changes in the operating temperature range.
ii. The thermal conductivity of the sample and reference material should be similar.
iii. The material should not react with the crucible or thermocouple.

Both methods provide similar information related to the characteristics of a material. The resemblance between the two methods includes the similarity between the DTA and DSC plots. Any thermal reaction will appear as a peak or deviation from the baseline, either in an upward or downward direction. The direction of the peak depends on whether it is an endothermic or exothermic reaction. However, there is a contrast in the representation of the same phenomena using both methods, as shown in Figure 4.14. In DTA, endothermic reactions are represented by peaks below the baseline owing to a lag in the sample temperature compared to the reference temperature. Whereas, for exothermic reactions, the sample temperature is higher and the resultant peak is shown above the baseline. However, in DSC, the convention adopted is opposite to that of DTA. The endothermic reactions are shown as peaks in the positive direction (above baseline) for DSC measurements. This is due to the requirement of a higher heat supply to the sample to maintain the same temperature as that of the reference. For instance, while DSC shows endothermic reactions such as melting as positive due to the requirement of a higher heat supply to the sample, DTA represents melting as a negative peak. Similarly, the crystallisation of a material is an exothermic reaction, which is represented as a negative peak in DSC but as a positive peak in DTA.

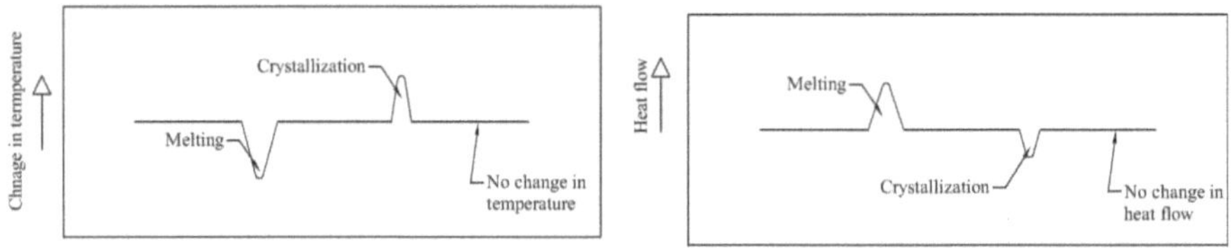

FIGURE 4.14 Typical differential thermal analysis and differential scanning calorimetry curves.

4.11.1 Differential Thermal Analysis

In DTA, the temperature difference between the sample and reference material, which are exposed to the same environment and heating rate, is measured as a function of temperature and/or time. A DTA instrument consists of a furnace, sample holder, temperature sensor, amplifier, temperature programmer, recorder and a control device for maintaining the desired atmosphere in the furnace, as shown in Figure 4.15. The temperature programmer controls the heat supply to the sample in the furnace. The instrument can typically be operated in a temperature range of

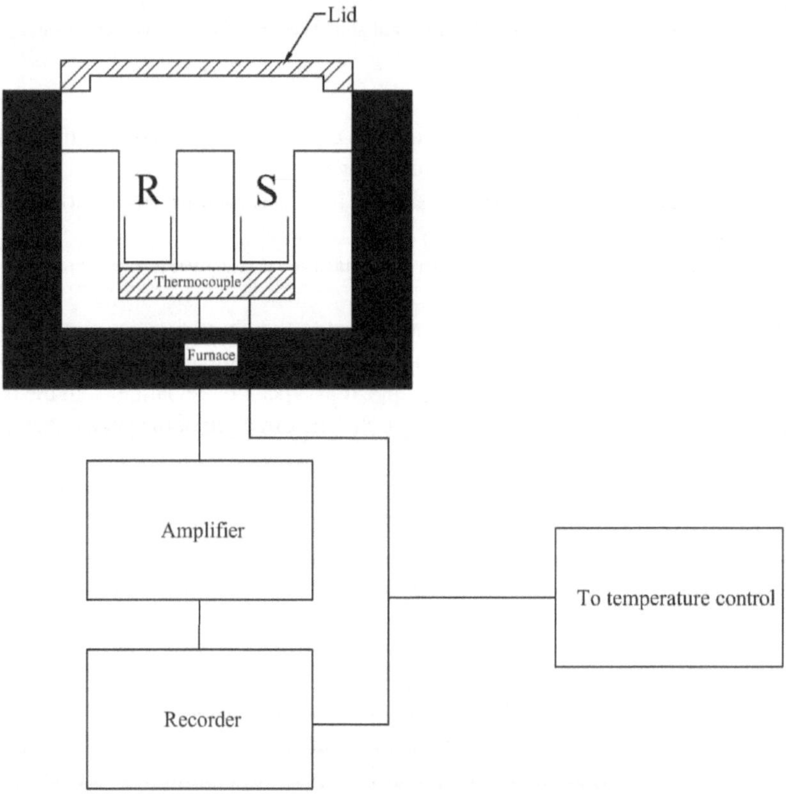

FIGURE 4.15 Components of the differential thermal analysis instrument.

−190°C to 2,400°C. In DTA, the samples are directly loaded into the sample holder inside the furnace without an additional pan for placing the samples. The thermocouples attached to the sample holder will generate signals corresponding to the temperature of the sample and reference, which are then fed to an amplifier. In the temperature range of −159°C to 250°C, a copper–constantan thermocouple is commonly used for measuring differential temperature, whereas for higher temperatures up to 2,400°C, tungsten–rhenium alloys and noble metals are preferred.

A DTA curve is a plot of the difference in temperature (ΔT) measured using a thermocouple as a function of temperature or time. The area under an endothermic or exothermic peak in the DTA curve is a measure of the enthalpy change for the particular thermal reaction. The endothermic reactions have peaks in the negative *y*-direction since such reactions are associated with the lowering of sample temperature. The temperature of the sample (T_S) will lag behind the temperature of the reference material (T_R) for endothermic reactions. For example, the melting of ice, which is an endothermic process, involves the release of heat during the phase transition from ice to water, thereby lowering the temperature of the sample. For an exothermic reactions such as oxidation, the peak is in the opposite direction (positive *y*-direction) since the sample temperature will be higher than the reference temperature. However, it is always preferred to mark the direction of endothermic or exothermic reaction in the DTA plot to avoid confusion. The change in temperature is calculated according to equation 4.4:

$$\Delta T = T_S - T_R \tag{4.4}$$

$\Delta T > 0$ implies an exothermic reaction
$\Delta T < 0$ implies an endothermic reaction

4.11.2 Differential Scanning Calorimetry

In DSC, the amount of heat or power to be supplied to the sample such that the temperature difference between the sample and the reference material is maintained at zero is measured. Both the sample and the reference are kept in the same environment and subjected to the same temperature programme. The amount of heat supplied is plotted as a function of temperature/time and the plot is called the DSC curve. There are two types of DSC instruments that are widely used: heat flux DSC and power-compensating DSC.

i. *Heat flux DSC:*

The heat flux DSC is a modification of the DTA, overcoming some of its limitations. In DSC, the sample is not in direct contact with the sample holder and thermocouple; instead, they are kept inside a crucible. Typically, the crucible is made of aluminium, platinum, copper or gold owing to their high thermal diffusivity. Moreover, the sample size used for DSC measurements is smaller than that for DTA measurements, which will help reduce the thermal lag in the measurements. The operating temperature range of a heat flux DSC is −180°C to 725°C in the case of an inert atmosphere and a maximum of 600°C for an oxidising atmosphere. The instrument measures the temperature difference between the sample and reference and calculates the heat flow based on the principle of the thermal equivalent of Ohm's law. The rate of heat flow is equal to the ratio of the temperature difference between the sample and reference to the thermal resistance offered by the sample, as given by equation 4.5:

$$dQ/dt = (\Delta T / R) \tag{4.5}$$

where dQ/dt is the heat flow rate, ΔT is the temperature difference between sample and reference, and R is the thermal resistance of the sample.

The sample and reference crucibles are placed on two separate raised platforms in a metal disc made of constantan, which offers a low-resistance path to heat flow. Constantan (55% copper and 45% nickel) is used because of its constant resistivity over a wide range of temperatures. The underside of the platforms is attached with a chromel disc to which chromel and alumel wires are connected. The resulting chromel–constantan thermocouple measures the differential heat flow between the sample and reference. Alumel wires attached to the chromel discs form another thermocouple to independently measure the sample and reference temperatures. Since the thermocouples are not directly in contact with the sample, the response to temperature changes is slower. But it has the advantage that the measurements are less affected by the thermal properties of the sample. All these components are enclosed in a single furnace, and the sample and reference are heated by a single heat source.

ii. *Power-compensating DSC:*

Unlike heat flux DSC, power-compensating DSC has two separate, identical sensors and furnaces for the sample and reference, as depicted in Figure 4.16. Owing to the double-furnace design, power-compensating DSCs are more sensitive to small phase transitions, provide more accurate values of heat capacity and enthalpy, and provide true isothermal performance. Sample holders made of the same material, typically platinum or

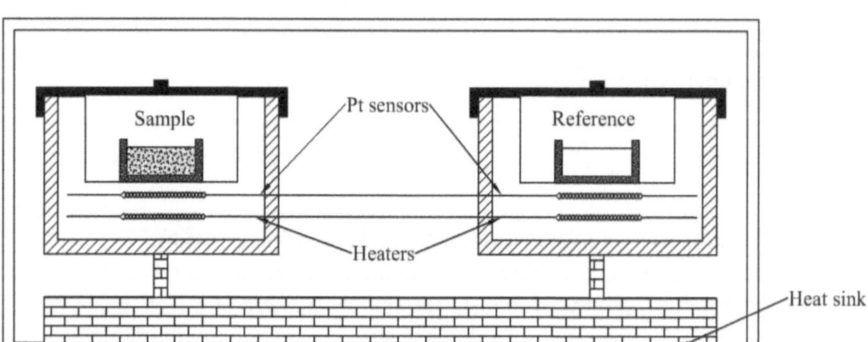

FIGURE 4.16 Power-compensating differential scanning calorimetry.

rhodium alloy, are used for samples and references. Samples are kept in a crucible, which is placed over the sample holder. The operating temperature range of a power-compensating DSC is in the same range as that of a heat flux DSC.

The power-compensating DSC instrument uses a feedback loop to maintain a similar sample and reference temperature while measuring the power or heat supplied to the furnaces to do it. It has two temperature control loops, one for average temperature control and the other for differential temperature control. Throughout the experiment, the sensors measure the temperature of the sample and reference, and the average of the output signals is compared with the programmer's signal. At all points in time, the average temperature (T_{avg}) measured and the programmed temperature should be maintained at the same level. The average temperature is given as shown in equation 4.6:

$$T_{avg} = T_S + T_R/2 \tag{4.6}$$

where T_S is the sample temperature and T_R is the reference temperature measured by the sensors.

The heat capacity of the holder with the sample will be generally higher than the reference due to the additional heat capacity of the sample placed in it. Hence, there will be a lag in the temperature of the sample compared to the reference. This temperature difference is recorded by the sensor, and the output signal of the sensor is amplified by a differential amplifier. Accordingly, the temperature programmer sends a signal to supply more power to the furnace whose temperature is lagging so that the imbalance in temperature between the two furnaces is corrected. This additional power supply will maintain the sample and reference temperatures at the same temperature. The differential power supplied is plotted as a function of programme temperature to obtain the DSC curve. If there is no chemical reaction or phase transition in the sample material, the differential power supplied will be equal to the heat capacity of the sample. Since it involves direct measurement of heat flow, it is also known as heat flow DSC.

Figure 4.17 shows the representative DSC plots of a sample obtained using the two types of DSC instruments. In a power-compensating DSC, the endothermic reaction appears as an upward peak as the instrument must supply more heat to maintain the sample and reference at the same temperature. On the contrary, in a heat flux DSC, the same reaction results in the absorption of heat by the sample, making it cooler than the reference and hence appearing as a downward peak.

4.11.3 Information from DTA and DSC Data

Thermodynamic properties such as enthalpy, entropy, Gibbs free energy, specific heat capacity, thermal conductivity, heat of fusion, heat of crystallisation, and kinetic data of materials subjected to temperature changes can be determined using DTA

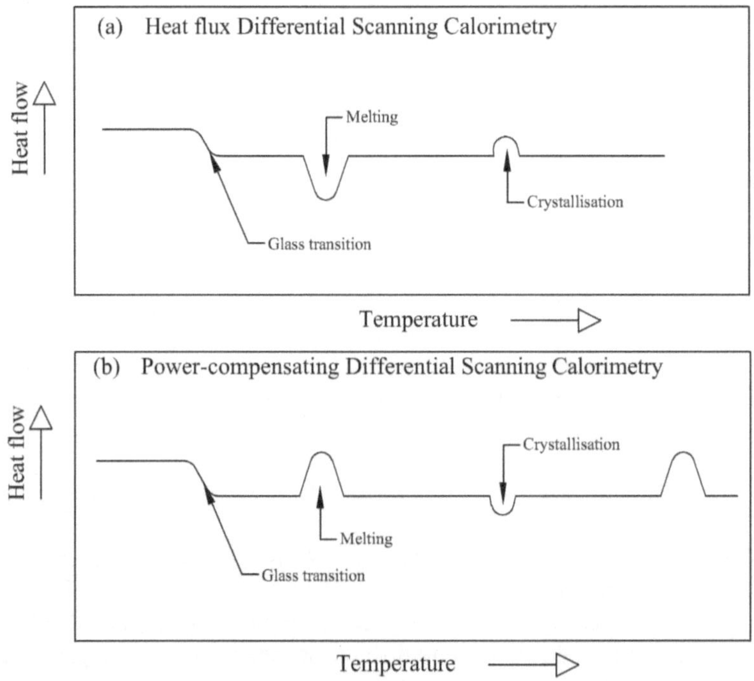

FIGURE 4.17 Comparison of heat flux differential scanning calorimetry (DSC) and power-compensating DSC plots.

and DSC. The determination of melting point, crystallisation point, glass transition temperature and purity of the substance is also possible with these methods.

4.11.3.1 Measurement of Melting Point

A typical endothermic peak corresponding to the melting of a pure substance measured using heat flux DSC is shown in Figure 4.18. To determine the melting point of a material, draw a tangent to the linear region of the rising portion of the peak and extend it to meet the baseline of the plot. The temperature at the point of intersection is the melting point of that material.

4.11.3.2 Measurement of Enthalpy Change

In a thermal equilibrium process, the change in enthalpy (ΔH) is equal to the heat (Q) added to or removed from the system. When there is a difference in temperature between the sample and reference, heat is supplied to maintain the isothermal condition between the two materials. The amount of heat supplied in DSC is used for calculating the enthalpy change. For processes that involve the absorption of energy (endothermic), the enthalpy of the system increases, whereas for exothermic processes, enthalpy of the system decreases. The absolute value of enthalpy is difficult to measure directly, and in practice, enthalpy change is calculated. The area under the curve representing an exothermic or endothermic reaction gives a measure of

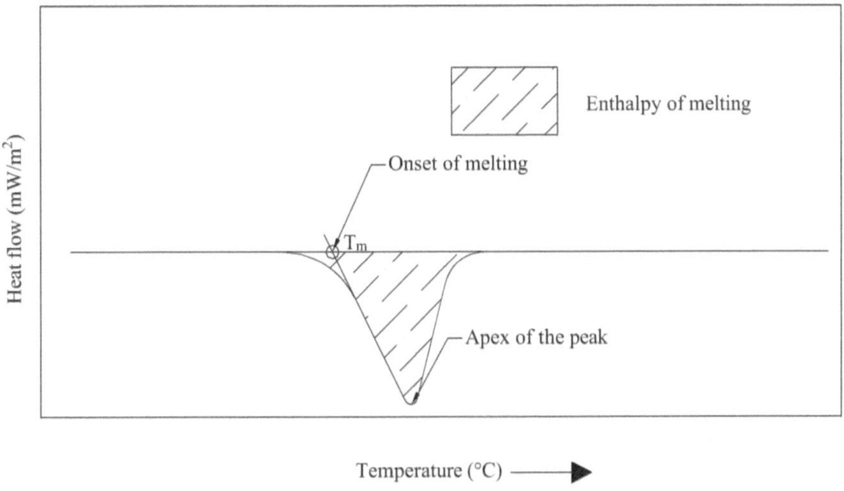

FIGURE 4.18 Differential scanning calorimetry curve for the melting of a pure sample.

enthalpy change in the sample. It is assumed that enthalpy change (ΔH) is proportional to the area of the peak (shaded region), as shown in Figure 4.18. The change in enthalpy is given in equation 4.7:

$$\Delta H = A \times K / m \tag{4.7}$$

where m is the mass of the sample, A is the area of the peak and K is the calibration factor.

The calibration factor is determined by relating a known enthalpy change, such as melting of a pure metal, to the measured peak area. Commonly, indium is used as the calibrant. The calibration coefficient is a constant for a particular instrument and testing conditions. For precise determination of enthalpy change, the testing conditions used for the actual sample and the calibrant should be the same. The temperature at which the measurement was taken has an effect on the calibration constant; hence, the calibrations should be performed over the entire operating temperature range of the actual experiment. The calibration factor is determined using the equation 4.8:

$$K = \Delta H_c \times m_c / A_c \tag{4.8}$$

where ΔH_c is the enthalpy change of the calibrant, m_c is the mass of the calibrant and A_c is the area of the peak corresponding to the calibrant.

4.11.3.3 Measurement of Heat Capacity

The heat capacity at constant pressure (C_p) of materials can be determined using DSC by measuring the change in enthalpy of the sample. The relation between enthalpy and heat capacity is given in equations 4.9 and 4.10:

$$\Delta H = \int_{T_1}^{T_2} C_p \, dT \tag{4.9}$$

or

$$C_p = \frac{\partial H}{\partial T} \tag{4.10}$$

Using DSC, only C_p can be determined since it is difficult to maintain a constant volume of the sample with variations in temperature. If heat capacity at constant volume (C_v) is required, then C_v is calculated from C_p using equation 4.11:

$$C_p - C_v = \left[\left(\frac{\partial H}{\partial T}\right)_T + p\right]\left(\frac{\partial V}{\partial T}\right)_p \tag{4.11}$$

4.11.3.4 Measurement of Thermal Conductivity

The thermal conductivity of a material can be measured using a DSC instrument. A cylindrical sample of 10–25 mm length is prepared for this purpose. Heat is supplied to the sample to reach a desired temperature, and once the output is steady with time, measure the temperature at both ends of the sample and the recorder displacement, h_s. The temperature at the bottom is obtained as the output of the DSC, and temperature at the top is measured using an additional thermocouple. The temperature difference across the sample is denoted as ΔT_s. The DSC instrument is calibrated with a sample of known thermal conductivity, for example, standard glass. The difference in temperature, ΔT_c, and displacement, h_c, is determined by replacing the sample with the calibrant and repeating the procedure. The thermal conductivity of the sample is given by equation 4.12:

$$\lambda_s = \frac{\lambda_c\left(h_s \times l_s \times d_c^{\,2} \times \Delta T_c\right)}{\left(h_c \times l_c \times d_s^{\,2} \times \Delta T_s\right)} \tag{4.12}$$

where λ_s and λ_c are the thermal conductivity of sample and calibrant, respectively; l_s and l_c are the length of sample and calibrant, respectively; and d_s and d_c are the diameter of sample and calibrant, respectively.

4.12 APPLICATION IN CONCRETE TECHNOLOGY

Applications of TGA include the characterisation of materials by analysis of decomposition patterns that are characteristic of the material, study of degradation mechanisms, understanding of reaction kinetics, determination of organic content in materials, etc. TGA is a widely applied technique to study the characteristics of hydrated cement paste. Since most of the hydration products, such as ettringite, monosulphoaluminate and portlandite, undergo decomposition or dehydration in different temperature ranges, thermal analysis can be used for the identification of the compounds present in the cement paste at different curing ages. The decomposition of cement hydrates can be generally divided into three stages. The first stage is associated with loss of water, and it can be further divided into two phases: evaporation of free water in the temperature range of 25°C–105°C and loss of bound water from the hydration products in the range of 105°C–400°C. The second stage corresponds to the decomposition of portlandite ($Ca(OH)_2$) to CaO and H_2O at

400°C–600°C. The third stage is associated with the decarbonation of $CaCO_3$ in the range of 600°C–800°C.

The calculation of weight loss due to the decomposition of various compounds in the paste helps in the quantification of the compounds. The weight loss peak corresponding to the decomposition of hydration products such as portlandite indicates the degree of hydration of the paste at different curing ages. As the curing age increases, the weight loss corresponding to the decomposition of portlandite increases. This implies that the hydration of the cement has progressed, resulting in higher quantities of portlandite in the hydrated cement. Similarly, the reactivity of pozzolanic materials such as fly ash can be assessed by studying the weight loss associated with portlandite. The amorphous silica present in the pozzolanic material will react with the portlandite available to form secondary C–S–H. A higher pozzolanic reactivity of the material will lead to faster consumption of portlandite and a reduction in the corresponding weight loss peak.

The durability properties, such as carbonation and sulphate attack, can also be studied by thermal analysis. TGA data can be used to study the extent of carbonation in the cement paste by measuring the amount of CO_2 released during the experiment. The possibility of sulphate attack can be confirmed by detecting the peak corresponding to the decomposition of brucite ($Mg(OH)_2$), which is a product of the reaction between $Ca(OH)_2$ and $MgSO_4$.

4.12.1 Identification of Compounds Present in Cementitious Systems

Table 4.1 shows the temperature range in which the thermal reaction of various cement hydration products and other compounds present in cementitious systems occurs. These compounds can be identified in an unknown cementitious sample by identifying the corresponding peaks in the thermogram of the sample. The thermal reactions associated with these compounds are described below:

i. *Gypsum:*

Gypsum is generally added to prevent the flash setting of the concrete. The addition of gypsum in sufficient quantities will slow down the rate of hydration of tricalcium aluminate (C_3A). In the absence of gypsum, C_3A present in cement reacts very rapidly with water to form calcium aluminate hydrate (equation 4.13). The initially formed hydrates (C_4AH_{13} and C_4AH_8) are in metastable states and further react to produce hydrogarnet (C_3AH_6), which is responsible for the quick setting of concrete (equation 4.14). The immediate stiffening results in less workable concrete and is mostly not preferred.

$$C_3A + 21 \rightarrow H\, C_4AH_{13} + C_4AH_8 \tag{4.13}$$

$$C_4AH_{13} + C_4AH_8 \rightarrow 2\, C_3AH_6 + 9\, H \tag{4.14}$$

In general, gypsum loses water in two stages as the temperature of the sample rises. At around 100°C–140°C, gypsum decomposes to form hemihydrate

TABLE 4.1
Temperature Range of Thermal Reaction of Different Hydration Products

Mineral	Characteristic Peak Temperature	Thermal Reaction
Kaolinite	Endothermic –450°C to 600°C	Dehydroxylation
	Exothermic –900°C to 1,000°C	Mullite or γ-alumina formation
Halloysite	Endothermic –100°C to 200°C	Dehydration
	Endothermic –450°C to 600°C	Dehydroxylation
	Exothermic –900°C to 1,000°C	Mullite or γ-alumina formation
Montmorillonite	Endothermic –100°C to 200°C	Dehydration
	Endothermic –600°C to 750°C	Dehydroxylation
	Exothermic –900°C to 1,000°C	Recrystallisation
Illite	Endothermic –100°C to 200°C	Dehydration
	Endothermic –600°C, 900°C to 920°C	Dehydroxylation
	Exothermic –920°C to 950°C	Recrystallisation
Vermiculite	Endothermic –100°C to 200°C	Dehydration
	Endothermic –800°C to 900°C	Dehydroxylation
	Exothermic –900°C to 1,000°C	Recrystallisation
Gibbsite and Goethite	Endothermic –250°C to 350°C	Dehydroxylation
Allophane	Endothermic –50°C to 150°C	Dehydration
	Exothermic –900°C to 1,000°C	γ-alumina formation
Quartz	Endothermic –573°C	α to β inversion

(equation 4.15), and it is later converted to anhydrite at 140°C–150°C (equation 4.16).

$$C\underline{S}H_2 \rightarrow C\underline{S}H_{0.5} + 1.5H_2O \quad (4.15)$$

$$C\underline{S}H_{0.5} \rightarrow C\underline{S}H + 0.5H_2O \quad (4.16)$$

The heating rate of the sample has a significant influence on the temperature at which the peaks corresponding to the dehydration of gypsum are observed. For higher heating rates, higher vapour pressure is created, which helps distinguish the two peaks better. Using crucibles with closed lids and a small opening also leads to well-defined peaks since the second conversion occurs at a higher temperature and the peaks are well separated.

ii. *Calcium silicate hydrates:*

Calcium silicate hydrate (C–S–H) is the main hydration product that contributes to the strength of a cementitious system. It is formed along with portlandite when tricalcium silicate (C_3S) (equation 4.17) or dicalcium silicate (C_2S) reacts with water (equation 4.18).

$$2\,C_3S + 11\,H \rightarrow C_3S_2H_8 + 3\,CH \quad (4.17)$$

$$2\,C_3S + 9H \rightarrow C_3S_2H_8 + CH \quad (4.18)$$

The weight loss related to C–S–H corresponds to the evaporation of interlayer water and the dehydroxylation of C–S–H. Besides, at a higher temperature of 800°C, the decomposition of C–S–H to wollastonite ($CaSiO_3$) is observed. Generally, C–S–H loses water in the wide temperature range of 50°C–600°C. Hence, TGA is not a good method for the identification and quantification of C–S–H. The amount of water lost from C–S–H depends on the sample preparation method adopted. Excessive and prolonged drying will affect the measurements since it leads to the loss of some quantities of bound water from C–S–H.

iii. *Portlandite and brucite:*

The dehydroxylation of portlandite is the most widely studied phenomenon in cement chemistry using TGA. The reaction involves the decomposition of $Ca(OH)_2$ into CaO and H_2O in the range of 400°C–500°C (equation 4.19). Corresponding to the decomposition, distinct peaks are observed in the DTG curve and are used for the quantification of portlandite produced. This is also indicative of the progress in the hydration of cement.

$$Ca(OH)_2 \rightarrow CaO + H_2O \tag{4.19}$$

Brucite ($Mg(OH)_2$) is formed along with gypsum when magnesium sulphate (MS) reacts with portlandite (CH) during an external sulphate attack of a cementitious system (equation 4.20). Magnesium sulphate may also react with C–S–H in the system to produce gypsum, brucite, and silica hydrate (equation 4.21). This results in a loss of strength in the cementitious system. The formation of silica hydrate indicates the degradation of C–S–H by the leaching of calcium from C–S–H.

$$CH + M\underline{S} + 2H \rightarrow C\underline{S}H_2 + MH \tag{4.20}$$

$$C\text{-}H\text{-}S + M\underline{S} \rightarrow C\underline{S}H_2 + MH{+}SH{+}H \tag{4.21}$$

At around 420°C, brucite decomposes into magnesium oxide (MgO) and water (equation 4.22). Hence, the detection of a weight loss peak associated with the dehydroxylation of brucite is used to study the sulphate attack of CH and C–S–H in cementitious systems.

$$Mg(OH)_2 \rightarrow MgO + H_2O \tag{4.22}$$

iv. *Ettringite and Thaumasite:*

In the presence of gypsum, C_3A reacts with water to produce ettringite ($C_6A\underline{S}_3H_{32}$) (equation 4.23). The formation of ettringite is found to slow down the reaction of C_3A by creating a diffusion barrier around it. Hence, the formation of ettringite in the early ages controls the stiffening of fresh concrete. Ettringite has a hexagonal prismatic shape with water molecules present on the surface of the crystal. The water from the crystal surface is lost at around 100°C, and the decomposition of aluminium hydroxide with the release of water occurs between 200°C and 400°C.

$$C_3A + 3\,C\underline{S}H_2 + 26\,H \rightarrow C_6A\underline{S}_3H_{32} \tag{4.23}$$

Thaumasite is formed when an external sulphate attack on concrete happens at low temperatures (<15°C) in the presence of carbonates in addition to sulphates. It has a structure similar to that of ettringite. Initially, it loses water at a slightly higher temperature of 130°C and continues to lose water up to 400°C. An additional peak observed in the case of thaumasite is associated with decarbonation, causing the release of CO_2 in the range of 600°C–750°C.

v. *Calcium monosulphoaluminates:*

Ettringite formed later reacts with the remaining C_3A once the gypsum present in the cement gets exhausted. Ettringite is then converted to calcium monosulphoaluminates. Due to this conversion, the diffusion barrier created by the formation of ettringite is broken, and more quantities of C_3A start to react (equation 4.24).

$$2\,C_3A + C_6A\underline{S}_3H_{32} + 4\,H \rightarrow 3\,C_4A\underline{S}H_{12} \tag{4.24}$$

AFm phases such as calcium monosulphoaluminates have a layered crystal structure with water molecules and anion interlayers in between the stacks of positively charged octahedral layers. The evaporation of interlayer water from monosulphoaluminates is observed in the temperature range of 250°C–350°C.

vi. *Carbonates:*

TGA of hydrated cement shows two or more peaks that are associated with carbonates present in the sample. Calcite, which is a major form of calcium carbonate, undergoes decarbonation in the temperature range of 600°C–800°C (equation 4.25). The release of CO_2 in this process is accompanied by weight loss, and a well-defined peak related to it is observed in the DTG curve. The amorphous form of calcite is also reported to decompose between 400°C and 600°C. In contrast, aragonite and vaterite, which are two other crystalline forms of calcium carbonate, recrystallise without showing any weight change at around 450°C and hence cannot be detected using TGA.

The presence of mono or hemicarbonates in cement also results in a weight loss peak at different temperatures. Firstly, five molecules of interlayer water present in the mono or hemicarbonates are lost from 60°C to 200°C, followed by the loss of six water molecules between 200°C and 300°C. A final peak at around 650°C can be observed due to the release of CO_2 from the carbonates. For a carbonated sample, the weight loss due to decarbonation in the temperature range of 600°C–650°C is indicative of the carbonation of hydration products such as portlandite and C–S–H.

$$CaCO_3 \rightarrow CaO + CO_2 \tag{4.25}$$

4.12.2 Quantification of Compounds in the Sample

Since the amount of mass loss corresponds to the quantity of the compound present, TGA can be used for quantitative analysis of samples. Quantification using TGA involves the determination of step heights of a thermogram. Each step denotes a change in weight corresponding to a thermal reaction such as decomposition, evaporation and oxidation. The sum of all the steps and the residue remaining after the last step will be equal to the weight of the sample before the start of the experiment. Sometimes, a few reactions may partially overlap, and the evaluation limits of each step required for quantification cannot be well defined. In such cases, taking the first derivative of the TG data and calculating the area under the peak of the DTG curve is more useful. DTG is a more visually acceptable form since the weight changes appear as distinct peaks. By measuring the area under the peaks, the absolute value or percentage change in weight related to the peak can be determined. Overlapping peaks observed in DTG curves can be resolved to a great extent by using numerical deconvolution methods. Deconvolution methods are used to separate complex thermal reactions so as to determine the relative area contribution of individual components (or reactions). It results in an accurate analysis of the sample, especially when the peaks are poorly resolved in the experimentally obtained DTG curve.

4.12.2.1 Methods of Quantification

Stepwise and tangential methods are the two methods that are mainly adopted for the quantification of samples using TGA.

i. *Stepwise method:*

   The stepwise method is most often used for quantification because of its simplicity. In this method, the weight change of the sample at the onset temperature (T_o) and the endset temperature (T_e) in the TGA curve is noted. The difference in weight change is a measure of the quantity of the compound (equation 4.26). The area under the peak of the DTG curve also gives the quantity of the compound. Figure 4.19 shows the quantification of $Ca(OH)_2$ using the stepwise method. To express the quantity in percentage,

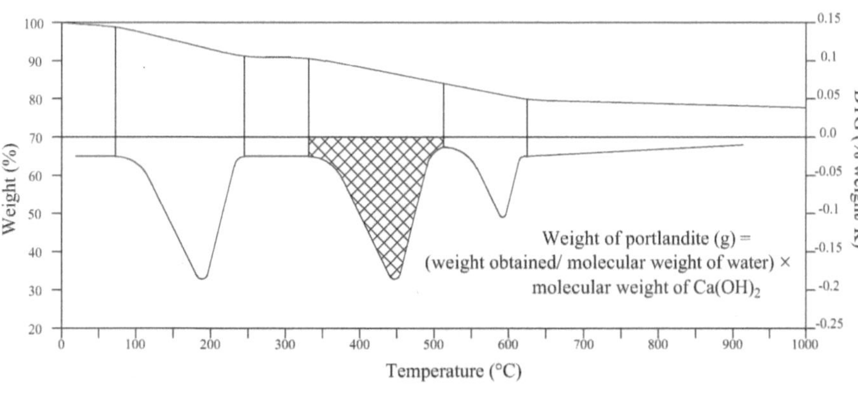

FIGURE 4.19 Quantification of $Ca(OH)_2$ using the stepwise method.

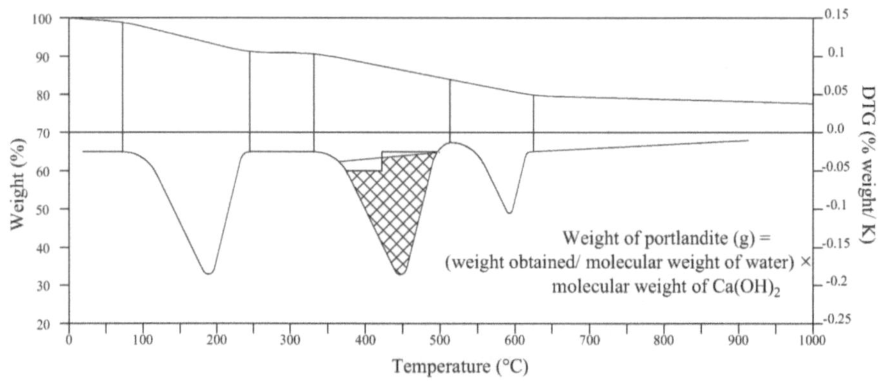

FIGURE 4.20 Quantification of $Ca(OH)_2$ using the tangential method.

the difference in weight change is divided by the initial weight of the sample at the start of the experiment and multiplied by 100.

$$\text{Quantity of compound} = \frac{m_o - m_e}{m_i} \times 100 \tag{4.26}$$

where m_e is the weight change corresponding to endset temperature, m_o is the weight change corresponding to onset temperature, and m_i is the initial weight of the sample.

ii. *Tangential method:*

In this method, tangents are drawn from the point of onset and endset of the DTG peak, as presented in Figure 4.20. Measuring the area under the tangents gives the quantity of the compound. The height difference between the two tangents drawn at the median temperature is calculated, even though in some software the difference between the tangents at the onset temperature is noted. A few other software are available that measure the difference in weight loss at the temperature where the maximum loss of weight occurs.

4.12.2.2 Quantification of Compounds Present in Cementitious Systems

i. *Quantification of bound water:*

The quantity of bound water present in the sample is indicative of the degree of hydration of cement. The bound water content is calculated by taking the difference between the weight loss of the sample at 105°C and 1,000°C. This calculation is adopted under the assumption that all the free water present in the pores is completely removed on heating up to 105°C. However, the loss of bound water from some of the hydration products, such as C–S–H, ettringite and monosulphoaluminate, under prolonged drying at a temperature below 105°C limits the applicability of this technique for quantifying bound water. To overcome this limitation, the weight loss corresponding to bound water can be obtained by measuring the difference in weight of the sample after solvent exchange and weight after heating up to

550°C. By taking the weight loss only up to 550°C, the weight loss region corresponding to decarbonation can be avoided.

ii. *Quantification of portlandite:*

The stepwise method is the most commonly used method to quantify portlandite in hydrated cement samples. In general, the dehydroxylation of portlandite occurs in the temperature range of 400°C–500°C. The weight loss corresponding to the onset temperature and endset temperature of the dehydroxylation peak is noted, and the difference gives the quantity of water evaporated from $Ca(OH)_2$. This weight loss due to evaporation can be used to obtain a measure of portlandite content in the sample, as shown in equation 4.27:

$$Ca(OH)_2 \text{ content} = WL \times \frac{M_{Ca(OH)_2}}{M_{H_2O}} = WL \times (74/18) \tag{4.27}$$

where WL is the weight loss corresponding to evaporation of water from $Ca(OH)_2$, $M_{Ca(OH)_2}$ is the molecular weight of portlandite (74 g/mol) and M_{H_2O} is the molecular weight of water (18 g/mol).

One of the limitations of the stepwise method is the overestimation of actual weight loss corresponding to portlandite since the weight loss calculated includes the weight of other compounds like C–S–H, which may lose weight in the same region as that of portlandite.

iii. *Quantification of calcite:*

Calcite decomposes into CaO and CO_2 in the temperature range of 600°C–800°C. Generally, most of the bound water from the hydration products will be removed at temperatures below 600°C. Therefore, the quantification of calcite is simpler compared to portlandite, as there is no overlapping of peaks. The procedure is similar to that followed for the calculation of portlandite content, as shown in equation 4.28:

$$CaCO_3 \text{ content} = WL \times \frac{M_{CaCO_3}}{M_{CO_2}} = WL \times (100/44) \tag{4.28}$$

where WL is the weight loss corresponding to release of CO_2 from $CaCO_3$, M_{CaCO_3} is the molecular weight of calcite (100 g/mol) and M_{CO_2} is the molecular weight of CO_2 (44 g/mol).

4.12.3 Compositional Analysis of Biomass

TGA can be used to determine the amount of volatile, medium volatile, and combustible matter and ash content in biomass such as sugarcane bagasse, rice husk and rice straw, which are excellent pozzolanic materials. For this, the weight loss of the sample is continuously recorded as it is gradually heated at a specific heating rate in different environments. When sugarcane bagasse samples are heated at a rate of 20°C/min with nitrogen as the purge gas, two main weight loss plateaus can be observed at 110°C (T_1) and 600°C (T_2) in the TGA curve, corresponding to the evaporation of highly volatile and medium-volatile matters, respectively. Once the second weight loss plateau is established, the furnace environment is changed from an inert to a reactive environment (with dry air as the purge gas), and the heating rate is lowered to 10°C/min. A third weight loss plateau can be observed at 800°C (T_3) due to combustion of the fixed carbon, and only inorganic ash remains after heating at this temperature. The various contents of the biomass can be calculated as shown in equations 4.29–4.32:

$$\text{Highly volatile content} = \left(\frac{w_o - w_1}{w_o}\right) \times 100 \tag{4.29}$$

$$\text{Medium volatile content} = \left(\frac{w_1 - w_2}{w_o}\right) \times 100 \tag{4.30}$$

$$\text{Combustible matter content} = \left(\frac{w_2 - w_3}{w_o}\right) \times 100 \tag{4.31}$$

$$\text{Ash content} = \left(\frac{w_3}{w_o}\right) \times 100 \tag{4.32}$$

where w_0 is the initial weight of the sample at the start of the experiment, and w_1, w_2 and w_3 are the weights of the sample at temperatures T_1, T_2, and T_3, respectively.

Figure 4.21 shows the thermogram and deconvolved DTG curves obtained upon heating a sugarcane bagasse sample at a rate of 10°C/min. It can be observed that the superposed, or cumulative fit, DTG curve is an exact fit to the experimentally obtained DTG curve. The deconvolution of the DTG curve of the bagasse sample helps to identify the lignocellulosic composition of the sample based on the temperature at which the peak corresponding to the weight loss is observed. For instance, weight loss due to hemicellulose decomposition in bagasse samples occurs at around

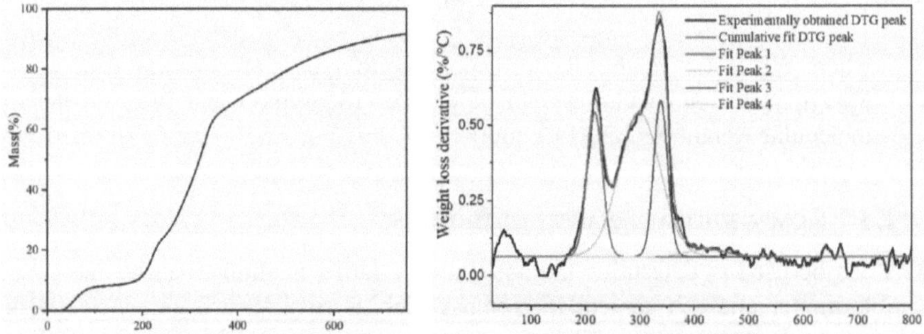

a) Thermogram

b) Deconvolved DTG curve

FIGURE 4.21 Thermogravimetric analysis data for a sugarcane bagasse sample.

300°C, which is the second peak after the peak corresponding to moisture loss. Similarly, weight loss due to cellulose decomposition occurs in the 315°C–400°C range. However, lignin is difficult to decompose, and the deconvolved curve for lignin is a broad, flattened peak ranging from 200°C to 600°C. The area under each of fitted curves is a measure of the relative amounts of the lignocellulosic compound present in the sample.

4.12.4 Application of DTA or DSC

The application of DTA or DSC in concrete technology is limited. This is due to the difficulty in the deconvolution of peaks corresponding to the reactions of different hydration products, which occurs in the same temperature range. Hence, these methods are not used for the quantitative analysis of hydration products. However, DTA and DSC can be used for the identification of MSH gel since it shows a characteristic peak at a temperature of around 850°C. This exothermic peak observed is related to the crystallisation of MSH gel into $MgSiO_3$. These methods can also be used for the characterisation of glass fractions in slag and fly ash. In DSC measurements for blast furnace slag, an endothermic peak corresponding to glass transition temperature (at 742°C), exothermic peaks indicating crystallisation (920°C–1,045°C), and endothermic peaks related to melting of some crystals (1,280°C–1,315°C) can be observed.

4.13 APPLICATION IN OTHER CIVIL ENGINEERING FIELDS

4.13.1 Pavement Materials

- TGA helps determine the thermal stability of asphalt binders. The asphalt starts to lose weight at 240°C, and maximum weight loss occurs around 460°C, which can be attributed to the thermal decomposition of asphalt. On heating the sample above 460°C, a horizontal plateau was observed in the TG curve, indicating continued weight loss at a lower rate. Finally, beyond 550°C, no significant mass loss was observed. The further breakdown of the asphalt was observed at around 600°C–620°C and 700°C due to the combustion of carbonaceous constituents.
- The glass transition temperature and crystallised fraction of the asphalt are temperature-dependent factors that influence the physical hardening of the asphalt cement. Therefore, DSC can be used to determine the glass transition temperature (T_g) and crystallisation of asphalt cement fractions. The DSC curve of the asphalt cement reveals three major thermal reactions: glass transition (–50°C to −10°C), small exothermic peak due to crystallisation (above T_g), and the broad endothermic peak corresponding to the dissolution of crystallised fraction (0°C–100°C).

4.13.2 Soil

- Various soil constituents undergo several thermal reactions on heating, such as oxidation of organic matter, oxidation of metallic ions, dehydration (evaporation of adsorbed water), dehydroxylation (removal of crystal lattice water) and recrystallisation.

- TGA, DTA and DSC techniques can be used for the identification of minerals in the soil. Table 4.2 shows the characteristic peaks in the DTA curve corresponding to major minerals in the soil and the associated thermal reactions.
- Quantitative determination of the mineralogical composition of the soil is possible using the DTA curve by measuring the height, area, or width of the main endothermic peak. However, for accurate quantification of the mineral, the size of the endothermic peak should be sufficiently large and

TABLE 4.2
DTA Endothermic and Exothermic Peak Temperature of Major Minerals in the Soil and the Thermal Reaction Resulting in the Peak

Compound	Temperature Range of Reaction	Thermal Reactions
Gypsum	100°C–140°C 140°C–150°C	Water is lost from the gypsum in two steps: gypsum to hemihydrate and hemihydrate to anhydrite
Portlandite	400°C–500°C	Dehydroxylation of $Ca(OH)_2$ into CaO and H_2O
Brucite	420°C	Dehydroxylation of $Mg(OH)_2$ into MgO and H_2O
Calcite	600°C–800°C	Decomposition of $CaCO_3$ into CaO and CO_2
Coarser calcite (from unhydrated cement)	700°C	Decarbonation of $CaCO_3$
Calcium silicate hydrate	50°C–600°C	Loss of interlayer water and dehydroxylation of C–S–H occurs over a wide range
Ettringite	100°C 200°C–400°C	Initially, water around the crystal is lost and from 200°C to 400°C dehydroxylation of AH_3 occurs
Thaumasite	130°C up to 400°C 600°C–750°C	Loss of water from the outer surface of the crystal, evaporation of hydroxides associated with silica and final peak corresponds to the loss of CO_2
Monocarbonate	60°C–200°C 200°C–300°C 650°C	Three peaks correspond to the loss of five interlayer water molecules followed by loss of 6 water molecules and evaporation of CO_2
Monosulphoaluminate	250°C–350°C	Interlayer water lost in three stages
Friedel's salt	140°C 250°C–400°C	Initially, interlayer water is lost and then main layer water is lost in two steps
Stratlingite	200°C	Loss of interlayer and main layer water
Calcium Aluminium hydrates	120°C	CAH_{10}
	270°C	Dehydroxylation of $Al(OH)_3$
	340°C	Hydrogarnets containing silica, C_3ASH_4
	260°C–320°C	Dehydroxylation of AH_3 and CAH
Magnesium silicate hydrate	270°C	Evaporation of bound water
	270°C–700°C	Dehydroxylation of the hydroxyl group
Hydrotalcite	270°C	Loss of four interlayer water
	400°C	Loss of six main layer water

well-developed. The estimation of minerals is also possible by measuring the characteristic weight change in the TG curve associated with the thermal reaction of a mineral. The thermal analysis techniques are usually combined with other techniques such as XRD or chemical analysis for higher precision and accuracy.

- The TGA is commonly used for determining the quantity of adsorbed water and crystal lattice water. The crystal lattice water for most of the silicate clay minerals evaporates in the temperature range of 150°C or 350°C–1,000°C.
- The thermal stability of organic matter in soil can be assessed by a thermal analysis of the soil. Moreover, the quality of organic matter can be characterised based on its thermal stability. An exothermic peak around 300°C–350°C is observed in the DSC curve, corresponding to the combustion of carbohydrates and other aliphatic compounds. A second peak associated with loss of aromatic compounds is detected in the temperature range of 400°C–450°C. In some cases, a third peak at a temperature between 500°C and 570°C is observed, corresponding to highly recalcitrant organic components.

BIBLIOGRAPHY

Bottom, R. (2008). "Thermogravimetric analysis. Principles and applications of thermal analysis". In: P. Gabbott (Eds.) *Principles and Applications of Thermal Analysis* (pp. 87–118). Blackwell Publishing Ltd, England. https://doi.org/10.1002/9780470697702.Ch3

Brown, M. E. (2002). *Introduction to Thermal Analysis: Techniques and Applications. Hot Topics in Thermal Analysis and Calorimetry* (Vol. 1). Springer, Dordrecht. https://doi.org/10.1007/0-306-48404-8

Loganathan, S., Valapa, R. B., Mishra, R. K., Pugazhenthi, G., & Thomas, S. (2017). "Thermogravimetric analysis for characterization of nanomaterials". In *Thermal and Rheological Measurement Techniques for Nanomaterials Characterization* (pp. 67–108). Elsevier, Amsterdam. https://doi.org/10.1016/B978-0-323-46139-9.00004-9

Menczel, J. D., & Prime, R. B. (2009). *Thermal Analysis of Polymers: Fundamentals and Applications.* John Wiley & Sons, New York. https://doi.org/10.1002/9780470423837

Ramachandran, V. S., Paroli, R. M., Beaudoin, J. J., & Delgado, A. H. (2002). *Handbook of Thermal Analysis of Construction Materials.* William Andrew, Norwich.

Wendlandt, W. W., & Gallagher, P. K. (1981). "Instrumentation". In: E. A. Turi (Eds.) *Thermal Characterization of Polymeric Materials* (pp. 1–90). Academic Press Inc, Cambridge. https://doi.org/10.1016/B978-0-12-703780-6.X5001-9

MULTIPLE CHOICE QUESTIONS

1. Which of the following phenomena are not associated with temperature variation?
   a. Sublimation
   b. Crystallisation
   c. Oxidation
   d. None of these

2. Match the following:

I - Differential Thermal Analysis	i - Measurement of heat flow or difference in power of sample and reference material
II - Thermogravimetric Analysis	ii - Measurement of temperature difference between sample and reference material
III - Differential Scanning Calorimetry	iii - Study of viscoelastic properties of a material
IV - Dynamic Mechanical Analysis	iv - Measurement of mechanical properties as a function of temperature
V - Thermomechanical Analysis	v - Measurement of mass loss associated with thermal reactions

   a. I - i, II - v, III - ii, IV - iii, V - iv
   b. I - ii, II - v, III - i, IV - iii, V - iv
   c. I - ii, II - v, III - i, IV - iv, V - iii
   d. I - i, II - v, III - ii, IV - iv, V - iii
3. Identify the phenomena that cause mass gain of the sample material?
      i. Oxidation;
      ii. Reduction;
      iii. Decomposition;
      iv. Dehydration;
      v. Absorption
   a. i, iv, v
   b. i and v
   c. ii, iii, iv
   d. i, iii, iv
4. What do you mean by quasistatic thermogravimetry?
   a. Sample is maintained at a static temperature for the entire duration of the weight loss measurement.
   b. The temperature of the sample is raised in several steps during the weight loss measurement.
   c. The sample temperature is maintained at a constant heating rate during the weight loss measurement.
   d. The sample temperature is initially increased to a value and then maintained constant during the weight loss measurement.
5. What is a thermogram?
   a. It is a plot of temperature difference of the reference and sample as a function of temperature of the sample.
   b. It is a plot between weight change and temperature of the sample.
   c. It is a plot between rate of change of sample weight and temperature of the sample.
   d. It is a plot between enthalpy change of the sample and temperature of the sample.

6. What is the type of reaction represented in the figure given below?

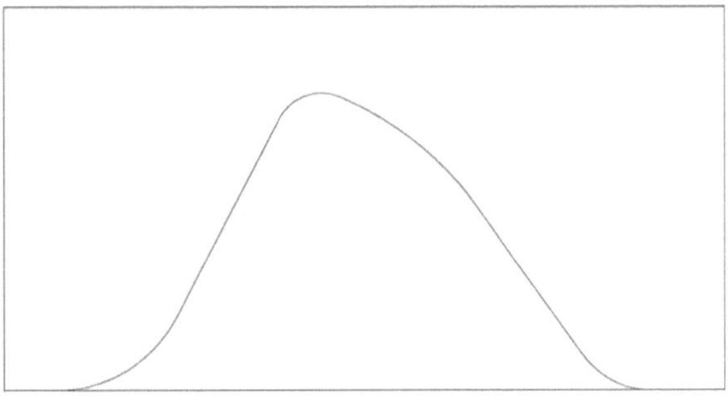

   a. Single-stage decomposition
   b. Crystallisation
   c. Oxidation followed by decomposition
   d. Melting
7. Which of the following information can be obtained from a thermogram of the sample?
   a. Thermal Stability
   b. Oxidative Stability
   c. Volatile content
   d. All of the above
8. What is the advantage of using derivative thermogravimetric curve?
   a. Helps to distinguish overlapped peaks of thermogram
   b. Area of DTG directly gives mass change of the sample
   c. Enthalpy of the material can be calculated
   d. Both a and b
9. Identify the following reactions marked in the DTA curve

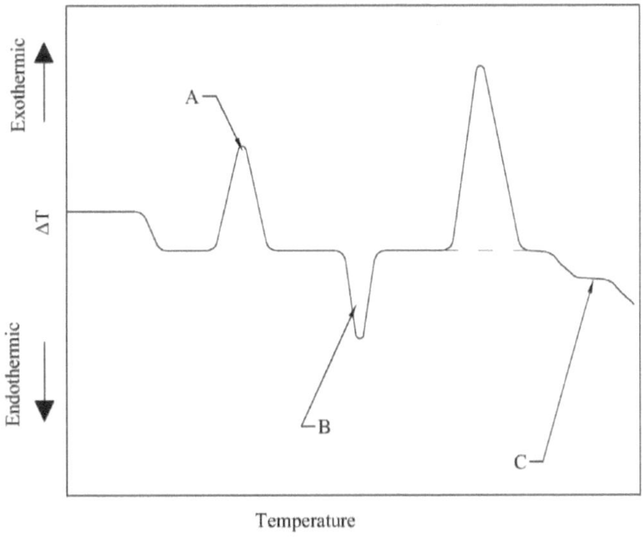

a. A-crystallisation, B-melting, C-Reduction
b. A-Melting, B-crystallisation, C-decomposition
c. A-crystallisation, B-melting, C-decomposition
d. A-Melting, B-crystallisation, C-reduction

10. Define enthalpy?
    a. Heat content of a system at constant pressure
    b. Measure of degree of randomness of a system
    c. Amount of heat supplied to the material at constant pressure
    d. Heat flow from a body with higher temperature to a body with lower temperature
11. Which of the following reaction is not associated with phase transition?
    a. Sublimation
    b. Crystallisation
    c. Oxidation
    d. None of the above
12. The difference between the actual sample temperature and the temperature measured by the sensor is known as ____.
    a. Phase lag
    b. Thermal lag
    c. Temperature gradient
    d. Reaction interval
13. What is zero line in a DSC curve?
    a. It is a flat line following a peak indicating no reaction.
    b. It is a line connecting two points in a DSC curve before and after the peak.
    c. It is the DSC curve plotted using measurements of empty crucible or empty instrument.
    d. It is the tangent passing through the starting point of a peak.
14. Which of the following statements are true related to analytical balance?
    a. Displacement of beam from its initial position is measured in null-type balance to calculate the mass change of the sample.
    b. Beam type and Helical spring type are null-type balances.
    c. Hooke's law is the basis of torsion-type balance.
    d. None of the above.
15. What is the purpose of plotting blank curve for TGA?
    a. For comparing thermogram of sample and reference material.
    b. Correction for buoyancy effect.
    c. To compensate for instrumental errors.
    d. To eliminate errors due to variation in furnace temperature.
16. What is the role of purge gas?
    a. To create a non-reactive atmosphere inside the furnace.
    b. Faster heat transfer to the sample.
    c. To remove gaseous reaction products that are evolved inside the furnace.
    d. All of the above.

17. **Statement A**: The purge gas inside the furnace is maintained static.
    Statement B: A gas stream flowing at a high rate may disturb the balance mechanism.
    a. Statement A is true and statement B is false
    b. Statement A is false and statement B is true
    c. Both the statements are true and B is the correct explanation for A
    d. Both the statements are true but B is not the correct explanation for A
18. Identify the correct statements related to crucibles.
    i. Alumina crucibles are preferred over platinum crucibles for temperatures above 640°C.
    ii. Platinum and copper crucibles may act as catalyst for certain reactions.
    iii. The rate of diffusion of volatile products into the surroundings is governed by the shape of the crucible.
    iv. Use of open crucibles leads to shifting of DTG peaks to a higher temperature.
    v. When the lid of crucible is kept closed during the experiment, the closely lying DTG peaks appear well separated.
    a. i, ii, iii and iv
    b. i, iv and v
    c. ii, iii and iv
    d. ii, iii and v
19. How does packing density of sample influence the measurement of mass loss?
    a. Denser packing leads to higher thermal conductivity and faster heating of the sample.
    b. Loose packing results in faster oxidation of the sample.
    c. Packing density doesn't have any effect on the heating rate of the sample.
    d. Both a and b.
20. Which of the following is not a limitation of solvent exchange method?
    a. Above 200°C, methanol decomposes to release CO_2 which causes carbonation of the sample.
    b. Prolonged immersion of sample in isopropanol affects the structure of ettringite and monosulphoaluminate.
    c. Only bound water can be removed using this method.
    d. All of the above.
21. What is the range of sample size used in TGA?
    a. 2–50 mg
    b. 2–50 g
    c. <2 mg
    d. Around 1 g
22. The mass loss of concrete sample in the range of 25°C–105°C corresponds to ______ and 400°C–600°C corresponds to _______.
    a. Evaporation of bound water, dehydration of portlandite
    b. Evaporation of free water, evaporation of bound water
    c. Evaporation of free water, decomposition of ettringite
    d. Evaporation of free water, dehydration of portlandite

23. Which of the following statements are true?
    a. A peak in the DTG curve of concrete in the range of 600°C–800°C indicates carbonation of concrete.
    b. The mass loss corresponding to portlandite is higher for a concrete cured for 91 days than for a concrete cured for 28 days.
    c. DTG peak corresponding to decomposition of portlandite is lower for blended concrete (cement + pozzolanic material) than OPC concrete.
    d. All of the above.
24. A mass loss peak at around 420°C indicates which durability problem of concrete?
    a. Carbonation
    b. Attack by $MgSO_4$
    c. Attack by chlorides
    d. Acid attack
25. Identify the compound: It loses water initially at 130°C and continues to lose water upto 400°C. Another peak in the range of 600°C–750°C is also observed due to decarbonation.
    a. Calcite
    b. Monocarbonates
    c. Thaumasite
    d. Aragonite
26. What is the temperature range at which the mass loss peak associated with recrystallisation of vaterite is observed in DTG curve?
    a. 400°C–600°C
    b. 600°C–800°C
    c. Above 800°C
    d. Does not show any mass loss
27. Which of the following cannot be quantified using TGA?
    a. Carbonate
    b. Portlandite
    c. Calcium Silicate Hydrate
    d. Bound water
28. Which of the following can be determined using DTA or DSC?
    a. Crystallisation point
    b. Melting point
    c. Glass transition temperature
    d. All of the above
29. What is the device used for measuring differential temperature?
    a. Thermostat
    b. Thermocouple
    c. Thermometer
    d. Infrared sensor

30. Which among the following are applications of DTA in cement industry?
    a. Identification of MSH
    b. Characterisation of glass fraction in fly ash
    c. Characterisation of glass fraction in slag
    d. All of the above

1 d	2 b	3 b	4 b	5 b	6 c	7 d	8 d	9 c	10a
11c	12b	13c	14a	15b	16d	17b	18d	19d	20c
21a	22d	23d	24b	25c	26d	27c	28d	29b	30d

5 Miscellaneous Characterisation Techniques

5.1 FOURIER-TRANSFORM INFRARED SPECTROSCOPY

5.1.1 Introduction

The study of interaction between electromagnetic (EM) waves and matter is called molecular spectroscopy. The above investigation can be used to study the structure of matter. The wide EM spectrum is considered to have radiations of different frequencies and wavelengths in simple harmonic wave propagation. These propagations travel in a straight line, unless disturbed to reflect and refract. A spectrum is created as a result of the manifestation of the disturbance produced by the interaction between matter and the propagated waves. All the waves in the EM spectrum travel like a sine wave. Consider a sine wave travelling with an angular velocity (ω) as well as a maximum amplitude of E (equations 5.1 and 5.2). Consider a point 'A' in the sine wave with an amplitude 'e', as shown in Figure 5.1.

If the point 'A' is moving in a circular motion, then it traverses an angle 'θ' in a time 't'. Therefore, the angular velocity is given by $\omega = \theta / t$. The displacement of the point 'A' in the vertical direction is given by Trigonometric laws. From Figure 5.2.

$$\text{Sin } \theta = \frac{e}{E} \tag{5.1}$$

$$e = E \text{ Sin } \theta \tag{5.2}$$

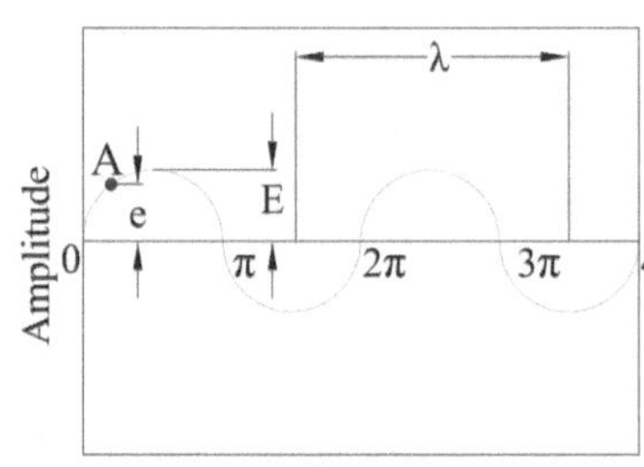

FIGURE 5.1 A sine wave with a wavelength λ and a point 'A' traversing the wave.

DOI: 10.1201/9781032635392-5

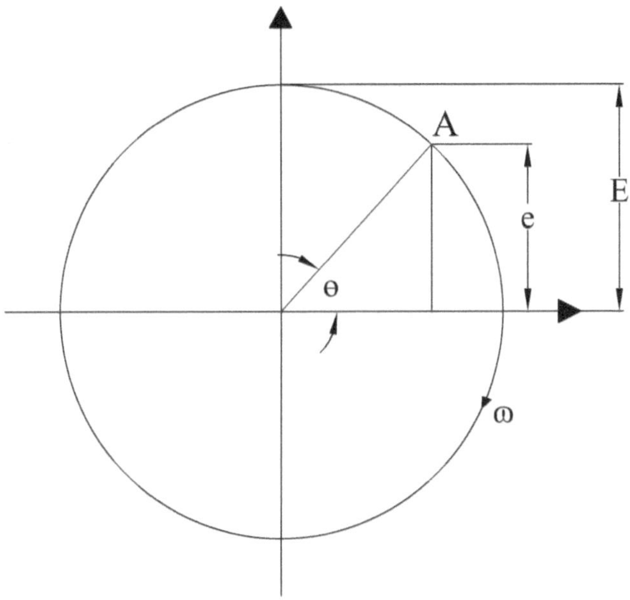

FIGURE 5.2 Motion of a point 'A' in a circle.

We know that

$$\omega = \frac{\theta}{t} \rightarrow \quad \theta = \omega t \tag{5.3}$$

Substituting equation 5.3 in equation 5.2 gives equation 5.4:

$$e = E \text{ Sin } \omega t \tag{5.4}$$

Now, consider the point 'A' traversed 2π in time 't', then the angular velocity,

$$\omega = \frac{2\pi}{t} \rightarrow f = \frac{\omega}{2\pi} \tag{5.5}$$

Substituting for ω in equation 5.4, from equation 5.5:

$$e = E \text{ Sin } 2\pi ft \tag{5.6}$$

We also know that the point 'A' travels with a linear velocity, 'v', the distance 'x' travelled by the wave is given by equation 5.7:

$$x = vt \tag{5.7}$$

Substituting equation 5.7 in equation 5.6

$$e = E \text{ Sin} \frac{2\pi fx}{v} \tag{5.8}$$

Equation 5.8 can be used to measure the distance travelled by the point '*A*' in a circular motion.

5.1.2 Concept of Energy Quantisation

Before 1900, it was thought that energy was taken continuously by matter, even though research by Max Plank shows that the energy is quantised. Hence, the energy is released or taken in the form of discrete packets at the sub-atomic level called photons. This discrete set of packets restricted to a set of prescribed values is called quantisation. As a result, based on the set of energy levels at which the electrons in an atom or molecule can exist, they can exist at different energy states. Subsequently, the rotational energy, electronic energy and vibrational energy of an atom or molecule are quantised. An atom or a molecule can exist at different energy levels by jumping from one energy level to another. Consider a system that has two energy states, E_1 and E_2, by exposing the system to an appropriate energy source ($E_1 - E_2$), the system can transit from one energy level to another. The energy difference is either absorbed or released in the form of EM radiation. The corresponding frequency can be calculated as shown in equation 5.9:

$$hf = E_1 - E_2 \tag{5.9}$$

where
$h = 6.63 \times 10^{-34}$ joules / molecule is the Plank's constant
f is the frequency.

The significance of the above equation is that if a system is exposed to monochromatic radiation (single frequency), the energy will be either absorbed or released from the system. The beam, after interaction with the system, is collected and analysed by a detector. If the intensity of the beam is decreased, then some part of the beam is absorbed by the system. The particular spectrum produced is called the absorption spectrum. In the same way, instead of monochromatic radiation, if an array of frequencies is used, only a specific frequency gets absorbed, and the rest will be emitted without any disturbance. Similar to the absorption spectrum, if the system emits energy and goes to a different energy state, instead of an absorption spectrum, the emission spectrum is obtained.

5.1.3 Spectral Regions

EM radiation is divided into different regions based on molecular processes. Table 5.1 shows the regions divided based on different energy levels, frequency levels and molecular processes. Interaction between the incident radiation and the matter results in the production of electrical or magnetic effects. The change in the

TABLE 5.1
Electromagnetic Spectrum Regions

Region	Wavelength	Frequency (Hz)	Molecular Process	Energy Level (J/mole)
Gamma ray	100–1 pm	10^{18}–3×10^{20}	Nuclear particle rearrangement	12×10^{9}–10^{11}
X-ray	10 nm–100 pm	3×10^{16}–3×10^{18}	Rearrangement of inner electrons	10^{4}
Visible and ultraviolet	1 µm–10 nm	3×10^{14}–3×10^{16}	Valence electron rearrangement	10^{5}–90^{5}
Infrared	100–1 µm	3×10^{12}–3×10^{14}	Vibrational spectroscopy	10^{4}
Microwave	1 cm–100 µm	3×10^{10}–3×10^{12}	Rotational spectroscopy	100
Radiofrequency	10 m–1 cm	3×10^{6}–3×10^{10}	Nucleus or electron spin reversal	0.001–10

electrical and magnetic fields of the radiation is a consequence of one of the following possibilities:

In the radiofrequency (RF) region, the spin of electrons and neutrons in an atom is accompanied by a magnetic dipole. If the spin of electrons and neutrons is reversed, then the interaction of the reversed spin with EM waves results in an absorption or emission spectrum. Similarly, if a molecule possesses a net positive or net negative charge, then it has a permanent dipole moment. Placement of the molecule with a net positive or negative charge on an EM field results in the rotation of the molecule around its centre of gravity. The rotation of the molecule is responsible for a dipole moment that changes periodically and fluctuates on a regular basis. The rotation of the molecule by reason of the absorption or emission of EM waves gives rise to a spectrum. For a molecule to be detected in a microwave (MW) region, it must possess a permanent dipole moment. If the molecule does not possess a dipole moment, then the net interaction with MW is zero. For that reason, the molecules are not detected in MW spectroscopy. In the case of the infrared (IR) region, the dipole moment caused by the vibration of the molecules is used to detect. The asymmetrical stretch or bending of vibration molecules results in a continuous change of dipole moment when radiating EM waves. This makes the molecule visible in the IR region. Similar to the IR region, the visible and ultraviolet (UV) regions make use of valence electron excitation, which consequently changes the dipole moments on interaction with EM waves.

5.1.4 Terminologies

Spectrum: A band of wavelengths produced by the refraction of white light to different degrees.

Absorption: Dissipation of light waves when passing through a material.

Emission: Production and discharge of EM waves.

Wavelength: is the distance between successive crests or troughs (refer to Figure 5.1)
Frequency: The rate of occurrence of EM waves in a stipulated time period.

5.1.5 Instrumentation and Spectral Representation

A schematic of a spectrometer is shown in Figure 5.3, and it is common for UV, visible and IR regions. It consists of an incandescent filament that emits radiation. Radiation emitted from the source passes through a slit and is reflected by a spherical mirror. After the reflection, a grating (reflective material block with ruled parallel lines on the surface) reflects the white parallel beam to a secondary spherical mirror. Reflection of the white source by a grating results in interference at the grid. This results in the reflection of different frequencies at different angles. The reflected frequencies from the grating pass through the spherical mirror and get focused on a exit slit. The radiation from the exit slit falls on a slot with a sample and gets focused on to a detector. For IR radiation, the detector will be a thermocouple. An amplifier amplifies the signal from the detector and records it as a spectrum that is collected and processed.

In case a sample of two energy levels is placed in the path of radiation. The detector detects the EM wave after the radiation passes through the sample. There will be a sudden drop in signal at a particular frequency. The drop at a particular frequency means that the sample has absorbed energy at that specific frequency that cannot reach the detector. A typical chart that shows the absorption of an energy level by the sample in the frequency spectrum is shown in Figure 5.4A. In reality, because of the limitation of the detector sensitivity and the dependence of the emissivity of the source on frequency, the baseline of the chart will never be horizontal, as shown in Figure 5.4B. Moreover, the slit cannot be made infinitely small, which leads to

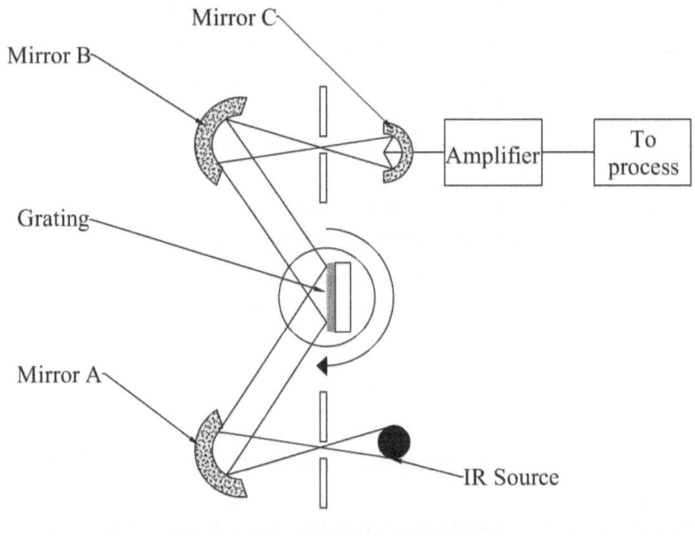

FIGURE 5.3 Schematic of a spectrometer.

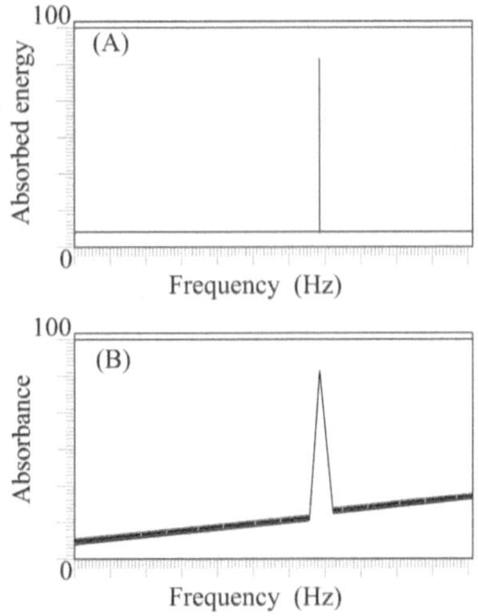

FIGURE 5.4 Single transition spectrum of a molecule. (a) Ideal. (b) Typical.

a range of frequencies detected by the detector. The transition of energy that takes place in an atom or molecule occurs over a range of frequencies. Based on its construction, the spectrometer can be classified as an absorption or emission spectrometer. Based on the topographical arrangement of the modulator, sample, analyser, detector and recorder, the absorption and emission spectrometers can be defined. In general, for an absorption spectrometer, a white or monochromatic source hits the sample and passes through a modulator, which is then scanned, analysed, detected and stored for processing. In the case of an emission spectrometer, the source light passes through a modulator, falls on the sample, and is then scanned by a scanning device, analysed, detected and then recorded for further processing.

5.1.6 Spectral Transitions: Width and Intensity

Spectral lines obtained during the processes of emission and absorption are not sharp but have broad peaks. A limited slit width in the spectrometer restricts the sharpness of the peak. Though the instrument's resolving power can be enhanced by improving the design, the inherent nature of the atomic and molecular scales restricts the natural line width. The restriction in the natural line width is a result of the imprecise determination of the energy levels of the system. Following are some of the factors that contribute to the fuzziness in the energy levels of the system: They are collision broadening, Doppler broadening and the Heisenberg uncertainty principle.

In the case of liquids and gases, the atoms are in continuous, random motion that results in deformation and disturbance of different energy levels. This disturbance affects the electrons in the outer shell resulting in a widening of the spectral lines. Similarly, spectral lines are disturbed because of rotational and vibrational spectra owing to the interference of collisions between the atoms. The separation of the atoms in the liquid is much smaller than the separation between the gases. This results in a less sharp peak for liquid samples compared to gaseous samples. On the other hand, for a solid sample, the extent and direction in which the particles can move are limited. This results in sharp peaks for solid samples. Doppler broadening results by reason of the Doppler Effect. The movement of particles in the sample results in a shift of frequency to low and high values. This results in the broadening of the peaks. The effect of Doppler broadening is the most predominant effect in the case of liquids. The uncertainty principle states that the particle's position and velocity cannot be determined precisely at the same time. Therefore, the sharpness of the energy levels is not infinite for a molecule or an atom. For any system, the energy level at its lowest state is defined as the most stable level. Eventually, the system will remain at that level for a longer time compared to the other levels. Nonetheless, the electrons in their excited state will be in that state for a maximum time period of 10^{-8} seconds. Hence, the magnitude of change in the energy level is 10^{-26} J. The corresponding radiation frequency is 10^8 Hz. Although the uncertainty in the RF is large, it is smaller when compared with transitions in the RF where the frequency is in the range of 10^{14}–10^{16} Hz. As a result, the effect of the uncertainty principle can be considered negligible for the RF region. On the other hand, it plays a vital role in the case of the excited electron spin state, where the frequency of transition is in the range of 10^8–10^9 Hz. In addition to the discussion on the width of the spectral lines, a brief understanding of the intensity of the spectral lines is also imperative. The intensity of the spectral line is influenced by the following factors. They are the probability of transition and population between energy states and the concentration of the sample. Energy states with a higher population have a higher probability of producing the most intense spectral lines. Statistically, this can be given by equation 5.10.

$$\frac{N_{\text{upper}}}{N_{\text{lower}}} = e^{(-\Delta E/kT)} \tag{5.10}$$

where

N_{upper} and N_{lower} are number of molecules in the upper and lower energy states:

$\Delta E = E_{\text{upper}} - E_{\text{upper}}$

k is Boltzmann's constant (1.38×10^{-23} J/K)

T is the temperature.

There exists a relationship between the intensity of the beam that passes through the sample and what gets absorbed or emitted from the sample, as shown in equation 5.11.

$$I = I_0 e^{(-kcl)} \tag{5.11}$$

where

I and I_0 are transmitted and incident intensity of radiation given by Beer–Lambert.

k is a constant specific to spectral transition.

c is the concentration of the sample.

l is the path length.

Transmittance and absorbance can be given as shown in equations 5.12 and 5.13, respectively:

$$\frac{I}{I_0} = e^{(-kcl)} = T \tag{5.12}$$

$$\frac{I_0}{I} = e^{(kcl)} = A \tag{5.13}$$

The ratio of transmitted and incident intensity, also called transmittance, is an exponential variation of the product of the spectral transition constant, sample concentration and path length. Similarly, the inverse of transmittance is called absorbance.

5.1.7 Fast Fourier-Transform: Conversion of the Time Domain to the Frequency Domain

In an IR spectrometer, an IR source is sent through a slit, which results in the generation of different sets of wavelengths. When the phase difference between the waves is constructive, then a certain set of wavelengths are produced that can go through the sample. The grating has its limitations in separating the wavelengths below a certain level. This results in the passage of certain sets of wavelengths through the sample. Thus, the passage of certain sets of wavelengths through the sample results in an interferogram.

The interferogram is a plot between the relative intensity of light and the path difference. Fast Fourier-transform does the conversion of the interferogram to an IR spectrum. Fourier-transform can be used to mention any function as a summation of sine and cosine waves. The specific wavelengths that correspond to the unknown function can be determined whenever an unknown function is represented in terms of sine and cosine waves. For that reason, a composite wave that consists of different frequencies is converted to separate waves whose frequencies can be identified or separated individually. These frequencies can be converted into the corresponding transition energies that take place in the bonds because of the vibration or bending of the atom or molecule. Eventually, a time-domain signal gets converted to a frequency-domain signal.

5.1.8 Applications

The bonds present in the system vibrate during the interaction between the molecules and EM waves. The vibration, in turn, results in the generation of peaks. Standardised peak values are available for vibrations that correspond to atomic or molecular combinations. The type of bonds and the elemental composition present in a compound can be identified by identifying the peaks during FTIR spectroscopy.

The asymmetric vibration of the Si–O bond in tri-calcium silicate (C_3S) and di-calcium silicate (C_2S) is observed as a dual peak at 938–883 cm^{-1} and 995–900 $cm^{-1,}$ respectively. Likewise, for tri-calcium aluminate (C_2A), the peak is at 898 cm^{-1}. In the same way, hydration products that are formed during the process of hydration can also be identified. The presence of portlandite ($Ca(OH)_2$) can be identified by a single peak at 3,640 cm^{-1}. The presence of Afwillite can be recognised by the asymmetric stretching of Si–O at 985–911 cm^{-1}. C–S–H gel can be detected at 1,000–950 cm^{-1}.

In some cases, decalcification of C–S–H results in a broadening of the peak from 1,100 to 900 cm^{-1}. Other hydration products such as ettringite and mono-sulfo-aluminate can also be determined based on the stretching vibration of O–H, S–O and Al–O bonds at 3,637, 1,115–617 and 571–537 cm^{-1}, respectively, for ettringite. In the case of mono-sulfo-aluminate, O–H, S–O, and Al–O bonds vibrate at 3,672, 1,150 and 579–525 cm^{-1}, respectively.

5.1.9 Major Advantages and Limitations

The major advantages of using FTIR are that it requires a very limited quantity of material, and the preparation of the sample is easier and less time-consuming. Attenuated total reflection mode can be used to study the surface of cementitious materials without any sampling or pre-conditioning methods. Although FTIR has the above advantages, it also has limitations, such as the fact that it cannot be used as a precise quantification method. Additionally, bands of trace elements such as calcium carbonate cannot be visualised properly. Their bands can be underestimated or erased because of the possible high intensity of other peaks. In addition, the presence of capillary water in the hydrated sample results in the disturbance of other peaks.

5.2 NUCLEAR MAGNETIC RESONANCE SPECTROSCOPY

5.2.1 Introduction

Atoms are considered to be the building blocks of elements. It consists of a central mass called the nucleus (positively charged) surrounded by negatively charged electrons. According to the Rutherford–Bhor model, the electrons revolve around the nucleus in an arbitrary path called an orbit. Classical mechanics states that neutrons, protons and electrons are considered elementary particles with a certain mass. Nonetheless, sub-atomic particles obey quantum mechanics laws. According to quantum mechanics, the relative probability density of an electron in a particular orbit in a specific space with a specific energy can be anticipated by Schrodinger's theory. Schrodinger's theory states that the relative probability density is a function of fundamental physical constants such as mass, the charge of the electron, radial distance or the co-ordinate axis of the electrons and the quantum numbers.

The quantum numbers that were introduced in the probability density function were considered to be absolute parameters in order to define the electronic wave functions of atoms. Primarily, four quantum numbers such as principal (n), azimuthal (l), magnetic (m) and spin (s) are considered in order to define the energy, size, shape, angular momentum, orbital direction, the effect of the magnetic field on the behaviour of electrons, and the electron's axial angular momentum. The principal quantum number governs the energy and the size of the orbital. The principal quantum number can take any integral value from one to infinity. The azimuthal or orbital quantum number also takes integral values from zero to infinity, although these values are less than the values of the principal quantum number. For example, for an n value of 4, the corresponding l values are 0, $1(n-3)$, $2(n-2)$ and $3(n-1)$. Based on the values of the azimuthal quantum number, the magnetic quantum number takes the values from ±l, ±(l–1) … 0. Therefore, for a l value of 3, the corresponding m values are ±3, ±2, ±1 and 0. The spin quantum number (s) takes the form of ½. The rotation about its own axis results in spin angular momentum. The spin angular momentum of an electron is given by equation 5.14:

$$\mathbf{s} = \sqrt{s(1+s)}\frac{h}{2\pi} \tag{5.14}$$

Substituting, for s in equation 5.14. The spin angular momentum is given by $\sqrt{3}/2 \cdot h/2\pi$ units. The spin angular momentum is a vector quantity that has a spin value of ±½. Similar to electrons, the nucleus rotates about its own axis. Due to the rotation of the nucleus about its axis, the nucleus also possesses spin angular momentum. Whenever the nucleus is placed in a magnetic field and excited at a specific angular frequency, the excitation produces a spectrum. This spectrum can be used to identify the composition of the material under consideration. Similar to electrons, the spin angular momentum of the nucleus is given by equation 5.15:

$$\mathbf{I} = \sqrt{I(1+I)}\frac{h}{2\pi} \tag{5.15}$$

The nuclear angular momentum can take zero, integral or half-integral values. Therefore, in the case of nuclear magnetic resonance (NMR) spectroscopy, the spin property of the nucleus is used in order to arrive at the chemical composition of the material under consideration.

5.2.2 Spin of the Nucleus: Effect of Applied Magnetic Field

Similar to the spin of the electron, the proton and the neutron in the nucleus also have a spin value of ½. For example, a hydrogen atom (^{1}H-Protium) consists of a proton in its nucleus and an electron held together by Columbic forces. The proton present in the nucleus accounts for a spin of ½. A deuterium (^{2}D) atom consists of a proton and a neutron; each of the sub-atomic particles in the nucleus undergoes a spin value of ½. The total spin value of the deuterium may be zero or one based on the spin, whether it is parallel or opposite in direction. Therefore, if a nucleus has '*a*' number of protons and '*b*' number of neutrons, then the total spin of the nucleus is the sum of the number of protons and neutrons in the nucleus, with each having a magnitude of ½. Each element in the periodic table consists of a different number of protons and neutrons in the nucleus. Eventually, their total spin number also varies from each other. Based on the odd and even numbers of the protons and neutrons present, the spin of the nucleus can be determined. The following are the simple steps to knowing the magnitude of the spin in different elements:

1. An even number of protons and neutrons in the nucleus leads to zero spin.
2. An odd number of protons and neutrons leads to an integral spin value.
3. An odd mass number (neutron + proton) leads to a half-integral spin value.

Spin in the nucleus leads to spin angular momentum, as given in equation 5.15. Angular momentum vectors point in some specific directions, with corresponding components equal to 2 I + 1. If the elements are in an unexcited state, all the components will have the same energy (i.e., they are degenerate in nature). On excitation by a magnetic field, different components are visible, and the degenerative state is removed.

Any charged particle rotating around its own axis possesses an electric current. Therefore, the charged particle acts as a magnetic dipole. The strength of the dipole is given, as shown in equation 5.16:

$$\mu = \frac{q}{2m} \quad I = \frac{qh}{4\pi m}\sqrt{I(1+I)} \tag{5.16}$$

where

q is the charge of the particle.

m is mass of the particle.

h is the Plank's constant.

I is the spin quantum number of the nucleus under consideration. The unit of magnetic moment is expressed in terms of Am^2.

Magnetic moment in terms of magnetic field strength is expressed in terms of tesla (T). Therefore, 1 Am^2 = 1 J/T. Substituting the magnitudes of q, m and h give 5.05×10^{-27} J/T. The magnetic dipole moment is given in equation 5.17:

$$\mu = \frac{q}{2m} \quad I = 5.05 \times 10^{-27}\sqrt{I(1+I)} \tag{5.17}$$

The value obtained by substituting the magnitudes of q, m and h is called a nuclear magneton, denoted as $\beta_N = qh/4\pi m$. In order to account for a factor for including the nucleus as a whole charge instead of a point charge, an experimental factor can be called a nuclear G factor. On accounting for the nuclear G factor, equation 5.17 becomes as shown in equation 5.18.

$$\mu = \frac{Gpeh}{4\pi m}\sqrt{I(1+I)} = g\beta_N I_z \tag{5.18}$$

where $g = Gp$, p is the number of protons and G is the nuclear factor. Based on the applied magnetic field, the dipole interaction varies, which results in the extension of the dipole component in a reference direction governed by I_z. Therefore, the applied magnetic field resolves I_z to different energy levels based on the net dipole moment experienced. The net dipole moment experienced at different energy levels can be separated by identifying the separation between different energy levels. The difference in energy levels can be given as shown in equations 5.19 and 5.20:

$$\Delta E = |g\beta_N B_z|\mathbf{Z} \tag{5.19}$$

$$\mathbf{Z} = I_z - I_{z-1} \tag{5.20}$$

In terms of frequency, equation 5.19 can be rewritten as shown in equation 5.21:

$$\frac{\Delta E}{h} = \left| \frac{g\beta_N B_z}{h} \right| \quad (5.21)$$

Therefore, if a nuclear spin transfers from one orientation to another, this results in the absorption or emission of energy in the form of radiation at a specific frequency. Moreover, the applied magnetic field and the radiation frequency are proportional to each other; by choosing and applying the appropriate magnetic field, the spin spectra of the nucleus can be studied. For example, for carbon 13 nuclei, on applying a magnetic field of strength 2.3487 T, emission or absorption of energy in the form of radiation occurs at a frequency of 25.14 MHz.

5.2.3 The Larmor Precession

The nuclear spin of a particular atom is considered to be analogous to the movement of a gyroscope. Similar to the precession of the gyroscope axis due to the application of a couple, the spin of the nucleus is also influenced and tends to take the direction of the magnetic field applied. The precession taken by the spinning nucleus to the applied magnetic field is called Larmor precession. The Larmor frequency is given as shown in equation 5.22:

$$\omega = \frac{\text{magnetic moment}}{\text{angular momentum}} = \frac{\mu B_z}{2\pi \mathbf{I}} = \frac{g\beta_N B_z}{h} \quad (5.22)$$

From the right-hand side (R.H.S.) of equations 5.21 and 5.22, it can be understood that the Larmor precession frequency is similar to the frequency separating two energy levels. Based on the above observation, it can be made clear that if the precession frequency of a nuclear spin in a specific atom matches the frequency of electromagnetic radiation, then a coherent interaction can be witnessed. This results in an exchange of energy between the EM waves and the particles. On the other hand, if the frequency is different, there may not be any interaction. This phenomenon is similar to resonance. If the above interaction takes place in the nucleus, then it is called NMR. Experimentally, this can be achieved by applying a constant frequency of radiation while varying the applied magnetic field until a transition occurs (absorption or emission) or vice versa.

5.2.4 Relaxation Times

Thermal motion consequently results in the distribution of nuclear magnetic moments in different energy states. The number or population of magnetic moments at different energy levels depends on temperature. In the absence of radiation, the nucleus orients itself and gets populated at different energy levels. The sharing of excess energy with the adjacent nuclei or the surrounding is called relaxation process. The time taken for the dissipation of excess energy is called relaxation time. Based on the relaxation process, two different relaxation times can be identified. They are longitudinal relaxation time (spin–lattice relaxation time) and transverse relaxation time (spin–spin relaxation time). In the case of longitudinal relaxation time, the spin energy is dissipated into the surrounding lattice by atomic vibration in the case of solids and by tumbling in the case of liquids and gases. Transverse relaxation occurs due to the sharing of excessive spin energy with the nearby nucleus. Measurement of relaxation times is important as the accuracy of the NMR spectral lines depends on the relaxation time. Based on the relaxation time, the NMR experiments can be broadly classified as broad-line (for solid samples) and high resolution (for liquid and gas samples).

5.2.5 Terminologies

Magnetic moment: The tendency of a magnetic object to orient in the direction of an applied magnetic field.

Angular moment: The rotational driving force developed in the course of a process.

Relaxation time: The time taken for the dissipation of excess energy is called relaxation time.

Resonance: Unified interaction due to the synchronisation in the frequency of EM radiation.

Precession: Change in the direction of the rotating axis of a body due to the applied torque.

5.2.6 Instrumentation and Techniques

An NMR system consists of an RF source, an RF detector, a sample holder, a recorder and a powerful magnet. Generally, the electromagnets that are used in an NMR require refined electronic circuits in order to steady the required current to produce a magnetic strength of 2.5 T. However, the corresponding frequency at which the NMR could operate is not more than 112.5 MHz. Therefore, in order to achieve higher magnetic strength, a powerful magnet capable of producing a wide range of frequencies is required. This can be achieved by 'superconducting magnets'. Superconducting magnets consist of liquid helium and gauge wires. As the gauge wires are in the liquid helium, the wires can provide a steady and large current without any loss, although the main disadvantage is that the work needs to be carried out at a low temperature and the expenses related to the system are reasonably high.

The sample to be tested is placed in an insulated cavity, and the electromagnets ensure the homogeneous distribution of the field over the sample. Once the sample is placed in the thermally insulated cavity, it is spun around in the field in order to obtain spectral lines of different nuclei in the sample. A typical schematic of the NMR spectrometer is shown in Figure 5.5. Based on the mode of observation, NMR is further classified as continuous wave NMR and Fourier-Transform (FT) NMR spectroscopy. Based on the nuclei under investigation, various quantities of samples are required. The sample holder consists of a long glass tube with a length of 15 cm and an external diameter of 0.5 cm. One end of the sample holder is closed, and the other end is open for filling the sample. The sample is filled to a length of 4–5 cm.

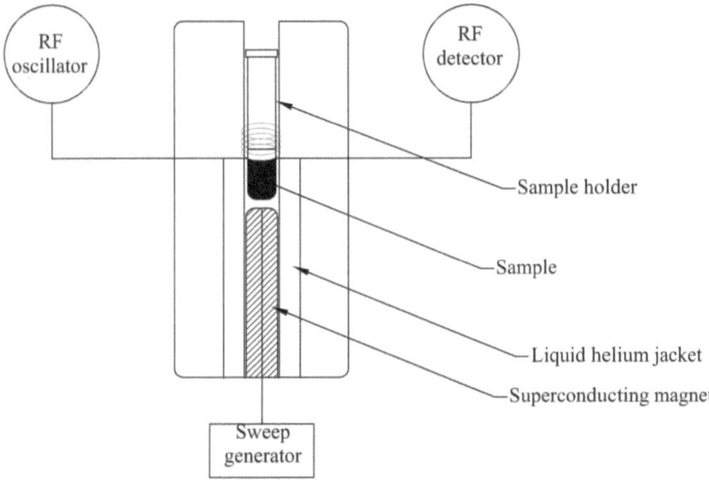

FIGURE 5.5 Schematic of a nuclear magnetic resonance spectrometer.

Usually, along with the sample, a small quantity of a standard such as tetra-methyl silane (TMS) is mixed, although if the sample could react with the reference, then the standard can be used separately in a tube.

Based on the requirement, either a field sweep or a frequency sweep can be applied. In general, a field sweep is easier to perform than a frequency sweep. During a field sweep, the coil (the source or the RF oscillator) placed around the sample radiates RF and sweeps the sample with radiation. The magnetic field is raised slowly and steadily by changing the corresponding current in the sweep generator. At different magnetic field strengths, different nuclei of the sample absorb the energy and resonate. After a certain time period, the nucleus in the excited state degenerates and comes back to the ground state. During this time period, the emitted energy is collected, amplified by the detector coil and passed to the detector. In some other cases, where double resonance experiments are carried out, the frequency sweep mode is used. In this case, the applied magnetic field is kept constant and the frequency is varied. To obtain a good spectrum, the quantity of the sample in the tube needs to be adjusted. In general, the higher the quantity, the better the spectrum. In an FT NMR spectrometer, the emitted signals are collected simultaneously by the detector as a function of time and stored for further analysis in a computer. The stored data in the time domain is mathematically converted into the frequency domain. A pulse of an appropriately wide frequency is swept across the sample. The frequency range needs to be wide so that all the hydrogen atoms in the sample get excited by absorbing the corresponding resonance frequency.

5.2.7 NMR for Hydrogen Nuclei

In most cases, NMR is conducted for a molecule in which more than one element is present. For example, in the case of a molecule with a hydrogen atom present in it, based on the paring element in the molecule, there will be a shift in the resonance frequency due to the influence of the electron clouds paired between the elements. This highly depends on the affinity of the pairing element to attract or repel the electron from the hydrogen atom. Therefore, when placing the molecules in the magnetic field, the effective field required to excite the hydrogen atom is the difference between the applied field and the induced field. As the induced field is proportional to the applied field, the effective field required to understand the behaviour of the nucleus is given as shown in equation 5.23:

$$B_{\text{eff}} = B_0(1-\sigma) \tag{5.23}$$

where σ is a constant and B_0 is the applied magnetic field. Therefore, the electrons surrounding the nucleus have a diamagnetic effect on the nucleus, which results in shielding the nucleus from the applied magnetic field. Consequently, the spectral line that corresponds to the hydrogen atom shifts based on the pairing elements in the molecule under investigation. For example, in order to understand the effect of pairing element on the NMR of a hydrogen atom, consider methanol (CH_3OH), in which a hydrogen atom shares a bond with oxygen in the case of OH and three hydrogen atoms share an electron with a carbon atom. Since oxygen has a higher electronegativity than carbon, that results in a lesser electron density around the hydrogen atom. Subsequently, the shielding provided by the electron cloud in the case of OH is less, which results in a considerably higher field experienced by the hydrogen nucleus than the field experienced by the hydrogen nucleus that is paired with carbon in CH_3. Eventually, the precession of the hydrogen nucleus bonded with oxygen takes place at a higher frequency than the hydrogen bonded with carbon. In contrast, a hydrogen atom with carbon requires a greater field to resonate (due to a higher electron cloud) than one with oxygen (due to a lesser electron cloud). This means that the hydrogen nucleus with oxygen has a smaller shielding constant, while the hydrogen nucleus with carbon atoms has a higher shielding constant. The hydrogen nucleus in OH resonates first, followed by the resonance of the hydrogen nucleus in CH_3. Since three hydrogen atoms are bonded with a carbon atom, the area of the resonance will be in a ratio of 3:1 when compared with OH.

Some of the important inferences from the above idea are that similar nuclei (here, hydrogen nuclei) resonate at different frequencies depending on the surrounding environment. This results in a shift in the resonance frequency occurring due to the chemical changes surrounding the environment, which is therefore called a chemical shift. Based on the number of nuclei paired together the area of the identified peak varies. The area is proportional to the number of equivalent nuclei. Besides the chemical shift, the field-induced circulation of electrons in the molecule may interfere with the resonance frequency. Two major prospects are due to the circulation of charge clouds in a cylindrical fashion and in an annular fashion. The two possibilities mentioned above may lead to de-shielding of the nucleus, thus resulting in resonance at low fields. Accounting for the chemical shift, the separation between energy levels is given in equation 5.24:

$$\frac{\Delta E}{h} = \frac{g\beta_N B_{\text{eff}}}{h} = \frac{g\beta_N B_0(1-\sigma_i)}{h} \tag{5.24}$$

The bare hydrogen nucleus is used as the primary standard to measure the chemical shift, although the shielding effect is the bare minimum, thus resulting in $\sigma = 0$. Thus, hydrogen atoms cannot be used as a standard, as their usage is not practical. Therefore, a universally accepted standard is necessary, TMS, with the following advantages: 1. a sharp, intense and equivalent resonance peak (all the hydrogen nuclei); 2. it is easily removable from other samples after NMR testing as it has a low boiling point; and it is used as a standard. The measurement scale is set in such a way that the resonance for TMS is set at 0 parts per million. All other resonances are measured with respect to TMS.

One more important technique that makes the characterisation by NMR more powerful is called the coupling constant. The coupling constant is based on the influence of the magnetic effect of one nucleus on the other. If the effect of the magnetic field of one nucleus (X) on another nucleus (Y) in an applied magnetic field (B_0) is of the magnitude B_X, then the line of force originating from one nucleus can influence the other nucleus in two ways. In the first case, B_X can be opposite to the applied magnetic field B_0, in this case, the total magnetic field experienced by the nucleus Y is $B_0 - B_X$. On the other hand, in the second case, B_X can be in the same direction to the applied magnetic field B_0, in this case, the total magnetic field experienced by the nucleus Y is $B_0 + B_X$. Consequently, the Y nucleus resonates at two different frequencies, forming a doublet with a separation of 2 B_X. Similar to the influence of X on Y, the influence of Y on X is also the same, forming a doublet. The effect of nucleus spin on the other nucleus is called coupling. Identifying the coupling constant is very useful in solid-state NMR. This can be used as a precision technique to measure the inter-atomic distance between different molecules in a compound. Until now, the discussion was based on the nuclear spin of a hydrogen atom; a brief discussion of the NMR of other nuclei is discussed in Section 5.2.8.

5.2.8 NMR for Other Nuclei

NMR can be obtained for any nucleus with a spin of ½. Other than the hydrogen nuclei, some of the important nuclei that can be detected and analysed in NMR are carbon-13, fluorine-19, phosphorous-31, silica-29, tin-117, tin-119, platinum-195, etc. Since the coupling effect and the chemical shift of the above-mentioned elements are in a higher range, their corresponding NMR spectra widely vary. For example, two phosphorous nuclei bonded to each other have a coupling difference of 600 Hz. Moreover, the tendency of the coupling to attenuate reduces as the bonds between the bonded nuclei increase. This results in practical advantages, such as the fact that the instrumentation that is required to determine the structural configuration requires less precision as the chemical shift that needs to be measured is larger. Therefore, the tolerance that is related to the measurement of the corresponding chemical shift is also larger. Besides, the NMR spectra obtained are less complicated.

For hydrogen, the NMR spectra vary from 0 to 10 ppm; similarly, for carbon-13, the spectra vary from 0 to 250 ppm. The abundance of carbon-13 nuclei is one percentage; apparently, the probability of finding a carbon-13 nucleus involved in direct coupling is very meagre. Eventually, the spin–spin coupling of the carbon sample can be ignored. On the other hand, a strong spectrum is visualised for a carbon-13 coupling with hydrogen nuclei. Based on the number of carbon atoms and the number of bonds between the carbon atoms, the coupling constant will vary. In general, carbon – 13 bonded to a single hydrogen atom gives rise to a doublet and carbon-13 bonded to two hydrogen atoms gives rise to a triplet. With three hydrogen atoms, a quartet is recorded. For a nucleus with spin values of more than ½, the transition induced by the energy is at the same frequency. Therefore, a single spectral line will appear for the nucleus. The position of the spectral line varies with respect to the chemical environment in which the nucleus is situated. Besides, based on the influence of the chemical environment surrounding the nucleus, the spectral lines split into a doublet or more. Although spectral lines are obtained for nuclei with spins greater than ½, they are not used much because of their intrinsic weakness and lack of clarity.

5.2.9 Major Advantages and Limitations

1. Liquid NMR sample preparation is easier and less time-consuming.
2. The spectral lines obtained by NMR spectroscopy could clearly indicate the type of the group, and the intensity of the spectral lines could provide the proportion of the groups.
3. The number of multiplets in the spectral line is very useful in identifying the number of protons in the group. In this way, the nearby neighbours of a molecule can be identified.
4. Although the NMR technique is versatile in nature, it cannot be used to detect groups that do not have magnetic nuclei. In those cases, an additional technique such as IR spectroscopy is used in order to locate the group.

5.3 ULTRAVIOLET–VISIBLE SPECTROSCOPY (UV–VIS)

5.3.1 Introduction

Similar to other spectroscopic techniques, UV–VIS works based on the Bohr–Einstein frequency relationship. UV–VIS works in the ultraviolet and visible regions of the EM spectrum. Both absorption and reflectance spectroscopy are possible in UV–VIS. In the visible range of the EM spectrum, the transition of atomic or molecular energy from one state to another is detected. The Bohr–Einstein frequency relationship is given by equation 5.25:

$$hf = E_1 - E_2 \tag{5.25}$$

where

$h = 6.63 \times 10^{-34}$ joules/molecule is Plank's constant

f is the frequency.

In terms of wave number, equation 5.25 can be given as equation 5.26:

$$h\bar{\lambda}c = E_1 - E_2 \tag{5.26}$$

where $\bar{\lambda}$ is the wavenumber and c is the velocity of the light.

The molecular or atomic transition results in a change in colour in the visible region. Subsequently, the change in colour during excitation can be used as a method of measurement. Moreover, the specific energy difference between the transitions can be linked to a specific frequency or the wavenumber of the EM spectrum. UV–VIS is used to excite the bonding and non-bonding electrons in an atom or molecule. The wavelength of UV–VIS ranges from 1 μm to 10 nm. Based on the bonding and non-bonding electrons, four types of transitions are possible. When electrons in the sigma bond and in the highest occupied molecular orbital transition to the lowest unoccupied molecular orbital, the transition is said to be a sigma–sigma transition. Similarly, the transition of an electron from a bonding pi orbital to an anti-bonding pi orbital is called a pi–pi transition. Likewise, a lone electron pair (i.e., electrons in an atom that can modify themselves in order to absorb light) denoted as n can have its own transition. These lone pairs of electrons have a transition from n to pi and n to sigma. These transitions of electrons from sigma to sigma, pi to pi, n to sigma and n to pi are four possible transitions in the case of electrons.

5.3.2 Terminologies

Spectrum: A band of wavelengths produced by the refraction of white light to different degrees.
Absorption: Dissipation of light waves when passing through a material.
Emission: Production and discharge of EM waves.
Wavelength: is the distance between successive crests or troughs (refer to Figure 5.1)
Frequency: The rate of occurrence of EM waves in a stipulated period.
Transmittance: The ability of an object or a material to allow light to pass through.

5.3.3 Background and Measurement Principle

UV–VIS works based on the Bouguer–Lambert–Beer law. This law is used to determine the intensity of light absorbed when it passes through the sample. The ratio of the light intensity that passes through the sample at a given wavelength (I) to that of the light intensity before passing through the sample at the same wavelength (I_0) is called transmittance (T). Transmittance can be used to measure the light absorbed by the sample. The light absorbed by the sample is given by equation 5.27:

$$A = \log_{10}\frac{I_0}{I} = \epsilon CD \tag{5.27}$$

where

ϵ is the excitation coefficient.
C is the concentration of the sample.
D is the sample path length.

According to the Bouguer–Lambert–Beer law, the light absorbed by the sample is directly proportional to the concentration of the sample under consideration as well as the width of the sample. Therefore, a calibration curve can be determined for a standard that gives the relationship between the change in absorbance and the concentration of the standard for a fixed length of the sample. The calibration curve can be used for the determination of the concentration of unknown samples. Based on the absorbance of a particular wavelength, the bond type and functional group in a molecule can be determined. The absorption spectrum of a compound can be established by obtaining a relationship between the wavenumber and excitation coefficient of the compound.

5.3.4 Instrumentation and Errors

UV–VIS consists of a light source made up of a tungsten lamp or deuterium lamp. The light passes through a prism or a diffraction grating in order to get a monochromatic beam. The monochromatic beam passes through a sample and a reference. Then it is detected by a detector and processed by a photodiode, multiplier tube or a charge-coupled device. Light before reaching the detector is scanned with a monochromator in order to allow only light of a single wavelength to reach the detector. The prism or diffraction grating is moved in a stepwise fashion in such a way that the intensity of each wavelength is measured. Many detectors are grouped together in an array fashion in order to collect and detect light of different wavelengths. Figure 5.6 shows the schematic of a UV–VIS.

Based on the beam split, UV–VIS can be a single- or double-beam spectrophotometer. In a single-beam UV–VIS, the light beam is passed through the sample. Subsequently, light intensity without any absorbance is measured by emptying or leaving the sample in the sample holder. On the other hand, a double-beam spectrophotometer consists of a beam split when it passes through the monochromator. At this point, the beam is split into two portions; one portion is passed through the sample, and the other portion passes through the reference. In the case of a double-beam spectrophotometer, the light beam that passes through the reference

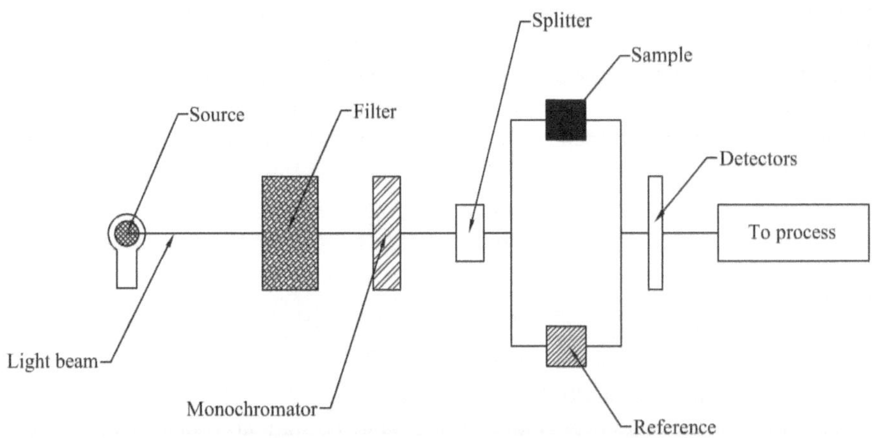

FIGURE 5.6 Schematic of an ultraviolet–visible spectroscopy (UV–VIS).

has an intensity of 100% transmittance (I_0). The measurement is given by the ratio between the light that passes through the sample and the light that passes through the reference. The sample that needs to be analysed is placed in a transparent cell called a cuvette. A cuvette is a cubic-shaped container with an internal dimension of 1 cm. A cuvette is made up of transparent materials such as fused silica or glass in order to allow ultraviolet and visible light to pass through. Although the cuvette used is transparent in nature, light reflects off the surface of the cuvette. A correction needs to be carried out in order to eliminate inaccuracies during the measurement. Subsequently, a blank reading is carried out with the cuvette being used in the experiment.

Limitations in the number of monochromators result in stray light during the experimental procedure, i.e., light with superimposed wavelengths. The scattered light from the surface of the monochromator results in this stray light. The proportion of stray light arising due to the scattering needs to be minimised or eliminated. Even though in most cases the stray light emitted is less than 0.1%, in cases where the wavelength of the light is less than 200 nm, the intensity of the absorbance light decreases due to conditions such as solvents or the surrounding air. This results in a reduction in the detection of peak intensity by the detectors. In the same way, in some spectral ranges of light, the intensity of the stray light is smaller than the intensity of the useful light. In those cases as well, some filters are used in order to filter out the stray light. The absorbance error due to the stray light can be corrected by the equation 5.28:

$$\Delta A = A' - A = \log \frac{T}{T(1-\alpha)+\alpha} \tag{5.28}$$

where

ΔA is the relative error in the absorbed light.

T is the transmittance.

α is the proportion of stray light. In general, the relative error in absorbance increases with respect to the rapid increase in the magnitude of absorbance for a fixed proportion of stray light. Subsequently, in order to minimise the effect of stray light, the absorbance of the sample needs to be reduced by choosing the appropriate concentration and path length.

5.3.5 Application

In the case of construction materials, UV–VIS is useful in the identification of the concentration of a specific material added to the compound. For example, absorptive performance can be used as a quantitative parameter to evaluate the percentage of asphalt present in oil specimens. Specifically, scanning spectra from 750 to 800 nm are considered ideal for identifying the solid content of emulsified asphalt in UV–VIS. Similarly, in the case of cementitious systems, admixtures are widely used for dispersion. The concentration of admixtures such as air-entraining admixtures and high-range water reducers can be determined by UV–VIS.

5.3.6 Major Limitations

Excitation coefficients obtained for a compound are independent of the substance concentration, although this is only appropriate if the solution is diluted. For concentrated solutions, the excitation coefficients depend on the refractive index (RI).

5.4 MERCURY INTRUSION POROSIMETRY (MIP)

5.4.1 Introduction

Construction materials are porous by nature. It is important to understand the porous nature of the material in order to enhance its engineering properties. Subsequently, different methods are available to determine the pore structure of the construction materials. Mercury intrusion porosimetry (MIP) is one of the methods used to understand pore size and pore size distribution. Especially when binders come into contact with water, they hydrate and during the process of hydration and concrete placement, voids with air are formed. These voids are classified based on their size and their characteristic occurrence. The voids formed during the placement process are typically mm to µm in size. Other types of voids include capillary voids and gel pores. The capillary voids are the spaces in which the hydration products or cement grains are not occupied. The pores in the gel of the hydration product calcium silicate hydrate (C–S–H) are called gel pores. The voids and pores formed during the process of hydration play a vital role in determining durability. As the ingress and the transportation of external agents will depend on the path and the connectivity of the pores. Subsequently, it is important to understand the pore size distribution and the connectivity of the binder system.

MIP involves using mercury (Hg) to intrude into the pores of the sample by applying pressure. Based on the contact angle between the Hg and the sample under consideration, the pressure required to intrude into the pores varies. In this section, background, measuring principles, experimentation, the major parameters that affect the measurement and their major limitations are discussed. Hg is considered to be extremely toxic in nature. Therefore, a short description of potential hazards and safety considerations during handling Hg is also given at the end of the section.

5.4.2 Terminologies

Intrusion: Forcible entry of a liquid (here Hg) into the voids of the solid.
Surface tension: The ability of the liquid surface to resist external force.
Wettability: The ability of the liquid to be in contact with the surface of a solid. Higher is the contact angle, lower is the wetting ability of the liquid on a surface and vice versa.
Contact angle: The angle between the solid surface and the liquid-vapour phase when they contact each other.
Capillary action: Flow of a liquid in a narrow pipe without any external force.
Porosimetry: A technique used to quantify the porous structure of a material.

5.4.3 Background and Measurement Principles

The MIP technique is used to determine the pore size distribution over a wide range of length scales, from a few nanometres (nm) to a millimetre (mm). Hg used for intrusion has a high surface tension, very low wettability or a high contact angle. In general, a contact angle of more than 90° is considered to non-wetting. Hg has a high surface tension; therefore, it does not spread on the surface. This results in less intrusion of Hg into any surface. Consequently, in order to penetrate a porous material, the non-wetting nature of Hg can be used. On applying pressure, Hg can penetrate porous material. Based on the magnitude of the applied pressure, the pore size of the material for investigation can be evaluated. The pressure that is applied during the experimental process is given by equation 5.29:

$$P = \gamma\left(\frac{1}{r} + \frac{1}{r_1}\right) \tag{5.29}$$

where γ is the mercury surface tension, r and r_1 are meniscus curvature radii. The curvature of the meniscus in porous media can be simplified, as shown in equation 5.30:

$$R = \frac{2\cos\theta}{r} \tag{5.30}$$

where

R is the radius of the curvature of the meniscus.

θ is the contact angle.

r is the radius of the capillary.

Combining equations 5.29 and 5.30, the Washburn equation can be obtained as shown in equation 5.31:

$$P = -\frac{2\gamma\cos\theta}{r} \tag{5.31}$$

The surface tension of Hg at 25°C is 0.485 N/m. From equation 5.31, pressure and the pore radii are inversely proportional. Consequently, increased pressure is required to intrude into a smaller pores. Similarly, an increase in contact angle increases the necessary pressure required to intrude on the sample. Although a contact angle of 140° is used in the case of a cement-based system, the contact angle depends upon various parameters such as chemical composition and impurities on the surface.

5.4.4 Instrumentation and Parameters

MIP consists of a filling station, a pressure unit, a dilatometer (sample container), etc. On the dilatometer, the sample is placed and degassed. Degasification removes the air from the air voids of the sample. Once the degasification is complete, the sample will be maintained in such a position that it will be floating in the Hg after it is filled with it. Now, the pressure is applied stepwise or continuously until the maximum pressure can be reached by the equipment. The volume of Hg intruding into the pore

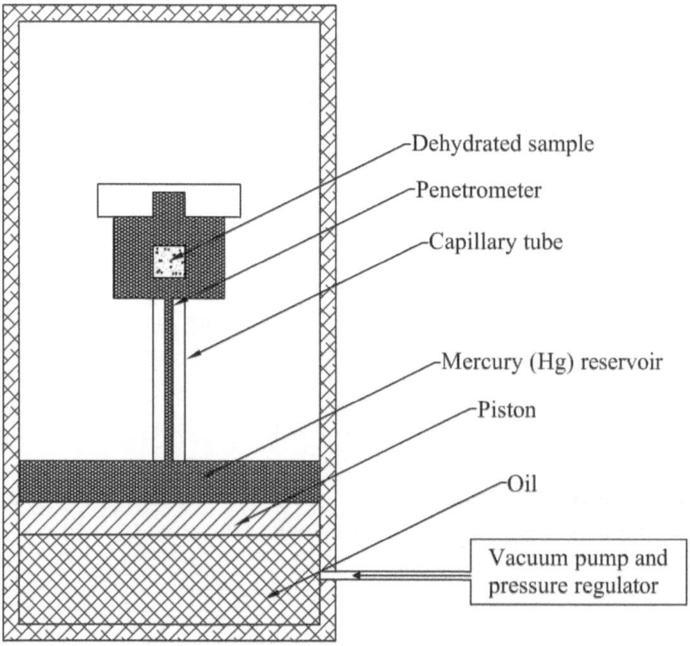

FIGURE 5.7 Schematic of mercury intrusion porosimetry.

of the sample is recorded. It is preferable to calibrate the equipment once a year with a standard such as porous glass beds. Once the calibration is over, a blank test is conducted without any sample in order to have a blank correction. The schematic of MIP is shown in Figure 5.7.

The pore size distribution of the sample can be calculated based on the intruded volume and the pressure of Hg, although some input parameters such as surface tension, angle of contact, mode and rate of pressure increment need to be provided manually in the corresponding software. The angle of contact depends upon many factors such as purity of the Hg being used, temperature, and roughness and chemistry of the surface. For a complex material such as cement, the contact angle varies from 120° to 150°. In pressurisation methods, especially in step mode, the pressure is increased in such a way that there is no further increase in the volume of the Hg intruding the sample. Two types are possible in the step mode: volume balance at incremental pressure and pressure balance at incremental volume. For most of the experiments, the step method is considered to be the best method for determining the pore size distribution because of the equilibrium achieved after each increment in pressure.

5.4.5 Representation of Results

MIP results are represented in terms of cumulative pore volume in the ordinate and pore entry radius in the abscissa. In general, the results are normalised with respect to sample volume or the weight of the material. Three important parameters can be derived from the MIP data. They are total percolated pore volume, threshold pore entry radius and critical pore entry radius. The total percolated pore volume is the total volume of connected pores in the system. It does not account for the unconnected pores and pores that are not intruded by MIP. The minimum pore entry radius that is continuous in the sample is called the threshold pore entry radius.

5.4.6 Influential Parameters and Their Effect

Increasing the angle of contact increases the pore size of the samples to larger values. The type of method used for pressurisation influences the pore size distribution, although the studies carried out to understand the influence of the pressurisation method on the pore size distribution are not conclusive. The rate of pressurisation is also considered one of the important parameters that can affect the pore size distribution. Based on the softness of the sample under investigation, the rate of pressurisation needs to be chosen appropriately. The size and weight of the sample influence the pore size measurement considerably. The bigger the specimen's size, the more reliable the measurement, as local heterogeneities in the sample can be avoided by choosing a larger sample. The too-big size of the sample can also have a relative underestimation of the pore size distribution due to the large ink-bottle effect. Subsequently, samples of similar mass with an increase in the number of pieces could reduce the boundary effect by increasing the surface-to-volume ratio.

During the sample preparation, samples are cut into regular or irregular shapes. Crushing the sample is not preferable as it may destroy the bigger voids, leading to erroneous results. It is important that the characteristics of the sample, such as the number and amount of the sample, are kept constant to get a reproducible result. The temperature of the room in which the experiment is carried out needs to be kept constant. Since variation in temperature changes the Hg density, that results in the wrong pore size distribution. Samples, especially cement-based materials, are prepared by various methods. Some of the methods used to prepare the samples are drying by oven or vacuum, freeze-drying and solvent exchange methods. Each preparation method has a different influence on the pore size measurement. For example, drying by heating results in a larger pore size distribution. Among the above methods, the solvent exchange method is considered to be the best method to retain the pore sizes of the cast sample without much difference during testing. The same method needs to be adopted for all the samples in order to have a comparison study between different samples.

The curing condition of cement-based materials has a major impact on the MIP test. For example, a sample cured in calcium hydroxide ($Ca(OH)_2$)-saturated water results in smaller pore sizes than a similar sample cured in an open atmosphere. This is due to the fact that curing the sample under water results in pore refinement. Similarly, an increase in the water-to-binder ratio increases the pore size of the sample and the total pore volume of the sample under investigation. An increase in the formation of hydration products results in increased densification of the cement matrix. Eventually, the pores in the matrix will refine. Refinement of pores results in decreased pore size distribution and less total pore volume.

5.4.7 Major Limitations

The following are some of the major limitations of MIP:

1. One major limitation of using MIP for pore size distribution is that the MIP cannot intrude through the unconnected pores.
2. Currently, the maximum allowable limit of MIP is 400 MPa, which corresponds to an intrusion of Hg at a pore size of 2 nm. Moreover, all the pores in the sample are considered cylindrical in shape. In reality, pores may not be cylindrical in shape.
3. MIP does not measure the real size of the pores. It can only measure the pore entry sizes. Therefore, the small pore size may act as an entry point for a large volume of the pore that may go undetected at low pressure. Since the entry point to the large-volume pore is via the small-sized pore. This effect is called the 'ink-bottle effect'. This results in an overestimation of small pores and an underestimation of large pores.
4. The pore size distribution detected by MIP is highly influenced by the input parameters. A change in the angle of contact given as an input parameter significantly influences the pore sizes determined by MIP.

5.4.8 Potential Hazards and Safety Requirements

Because of the highly toxic nature of Hg, 147 countries worldwide have banned its usage in any form. Hg is a volatile, liquid metal with no odour or colour in its vapour form. As Hg is very volatile in nature, even at a temperature of 20°C, care should be taken to avoid any spillage. Especially, breaking Hg into small droplets needs to be avoided at any cost since, the fragmentation of Hg increases the surface area exposed to the atmosphere. This increases the volatility rate. The use of safety gloves, glasses and masks is recommended when working extensively with Hg. Exposure to Hg does not immediately affect health; long-term exposure results in severe to very severe damage to the kidneys, lungs and nervous system. Breathing trouble may result from the exposure to a high concentration of Hg.

The room in which the MIP is placed should be isolated. Moreover, the Hg vapour is highly toxic in nature; it is required to maintain the air ventilation of the room. It is important that, after each test, the room be aerated and well ventilated. Maintaining the room temperature at a constant value of 20°C is highly important, as the increase in temperature increases the vapour release. Hg can combine with other metals chemically; therefore, it is important that when working with Hg, no metal-related objects should be worn. Hg should not be poured directly into any other container for disposal or cleaning purposes. Personal care products worn during the process of cleaning need to be handled with care as they may contain portions of Hg sticking to them.

5.5 PARTICLE SIZE ANALYSIS BY LASER DIFFRACTION

5.5.1 Introduction

Particle size distribution (PSD) is an important attribute in order to understand the physical properties of the material. The significance of determining the PSD is relevant in most fields, not just civil engineering. The size of the particle predominantly influences the fresh and hardened properties of the material. For example, the PSD and shape of the particles determine the surface area, which in turn influences the reactivity of the system. The packing of particles in a system depends on the PSD. For a system with mono-sized particles, the shape of the particle determines the packing density, while in the case of a multi-disperse system, PSD plays a pivotal role. PSD can be determined by many methods. Especially for a system in which the particles are in the micron and sub-micron ranges, the following techniques are predominantly used. They are diffraction (static scattering), obscuration, high-definition image processing, dynamic light scattering (DLS), nano-particle tracking analysis, etc. DLS is one of the most commonly used techniques for the analysis of PSD in the mono-dispersed system. The fluctuations in the light scattering caused by the Brownian motion of the particle are used to determine the size of the particle. Although DLS is used as an industrial standard for particle size measurement, the resolution of DLS is poor when measuring PSD in a multi-disperse system. Due to the above limitation, for a broader range of particle size measurement, static light scattering or laser diffraction (LD) techniques are widely used.

The LD technique involves irradiating a diluted suspension with a laser. The light scattered during the process is focused on a photodetector. The photo detector processes the scattered light. Based on the angle of scattering and the intensity of the light, the size of the particle in the system can be determined. Scattering models by Fraunhofer or Mie are used to arrive at the PSD. While the Fraunhofer model does not require the RI of the system, the Mie model requires the RI of the particle and the dispersing medium under examination.

5.5.2 Laser Diffraction Spectrometry (LDS)

LDS is an instrument that uses the diffraction pattern in order to measure the size of the particles. According to diffraction theories, the size of the particle is directly proportional to the scattered light intensity. In the same way, the size of the particle has an inverse relationship with respect to the scattered beam angle. LDS consists of a laser source (usually a helium–neon source), a laser tube, a power supply, etc. Light energy from the laser is passed through the suspension and detected by sensors. A lens focuses diffracted light after it passes through the suspension on the detector. The focal length of the lens determines the size of the particles the laser can analyse. The particle size that can be measured is directly proportional to the focal length of the lens used. Once the detector detects the diffraction, the beam is analysed for particle size and shape. Figure 5.8 shows the schematic of a typical LDS.

Diffraction of light waves takes place when the light passes through an obstacle or an opening. Obstruction of light waves around the corner of a particle (in the case

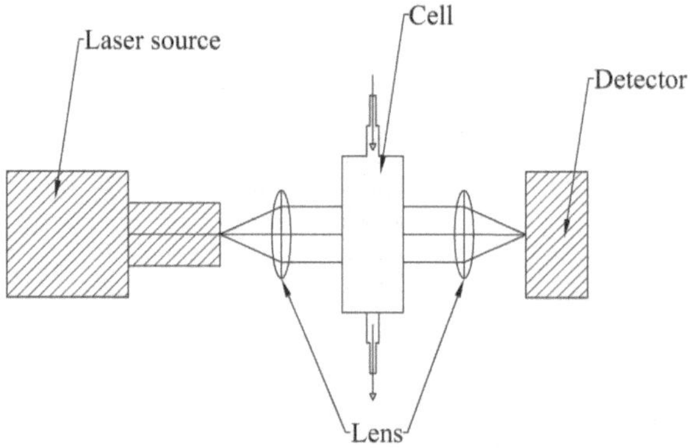

FIGURE 5.8 Schematic of a laser diffraction spectrometry.

of LD) results in the bending of the waves. The bending of the waves is independent of the composition or refractive properties (RI) of the material. Once the laser passes through the dry dispersion or the diluted dispersion, the light waves get absorbed, scattered and diffracted. Absorption takes place due to the attenuation of the light intensity. As the laser passes through the material, the light energy gets converted to other forms of energy, such as heat. The absorption of light is governed by the size and composition of the material. A certain portion of the light scatters, gets reflected and is refracted. Reflected and refracted waves are highly influenced by the size and composition of the material under investigation. PSD in LDS can be conducted in dry as well as wet states. In both cases, it is important to use appropriate refractive properties in order to get a precise measure of PSD. In LDS, a diffraction pattern obtained for a specific material is captured. The captured pattern is matched with the calculated pattern; the calculated patterns are obtained by Mie theory or Fraunhofer theory.

5.5.3 Terminologies

Reflection: Deviation of light waves due to the obstruction of the waves by a body (waves return back to the same medium).

Refraction: Propagation of light waves into different mediums due to the deviation of light waves.

Diffraction: Formation of a fringe due to the interference of light waves.

Scattering: Deflection of light waves.

Laser: Monochromatic light source generated by amplification and stimulated emission of radiation.

Absorption: Dissipation of light waves as different energy, such as heat, when the light passes through the sample.

Transparent: A material's ability to allow light to pass through.

Agglomeration: Collection of a number of particles due to physical forces such as van der Waals forces.

5.5.4 Measurement Techniques and Theories

Two different techniques can be used to measure PSD in LDS. They are wet and dry techniques. Based on the practical requirement, a choice of measurement can be made. In the case of the wet technique, the material for which the PSD needs to be determined is dispersed in a suitable liquid. The liquid used should be non-reactive in nature. The next important step is to disperse the particles in the liquid by means of mechanical agitation or sonication. In some cases, dispersants are also used to disperse the particles. The time of mechanical agitation or the percentage of dispersant added depends on the requirement. The light passes through the diluted suspension and undergoes absorption, scattering and diffraction. Based on the refractive properties of the material under investigation, absorption, the angle of reflected and refracted waves changes. Therefore, it is important to choose appropriate refractive properties for the materials that are considered for the determination of PSD. In the dry technique, the powder is directly fed into a stream of compressed air or vacuum; the powder gets dispersed and passes through the laser for the determination of PSD.

In the case of construction materials, both dry and wet methods are used. In view of practicality, the binders are dispersed in a liquid medium during the process of casting. It is appropriate to dilute the binders in a suitable liquid for analysis. Cement, as and when it comes in contact with water, starts reacting, resulting in a change in the size of the particles. Consequently, a non-reactive liquid is used. In most cases, isopropyl alcohol is selected. PSD of the material measured in LDS is not independent of the composition of the material. Most of the construction materials are multiphasic in nature. Subsequently, their surface properties have a considerable influence on the measurement of PSD. Moreover, the analysis of light waves after they pass through the sample requires the most suitable method. In those cases, it is important to use suitable refractive properties of the material in order to obtain reliable PSD.

The detectors detect the diffracted waves from the sample, and the diffracted waves are analysed by applying different theories. For particles a few micrometres in size, the Fraunhofer theory can be used. Usage of Fraunhofer theory to analyse PSD does not require the refractive properties of the material under investigation. Similarly, the Mie theory is used for finer particles. The use of Mie theory requires RI of the materials. Since Mie theory considers reflected and refracted parts of the light after passing through the sample. RI comprises a real part and an imaginary part. The real part depends on the nature of the material. At the same time, the imaginary part is related to the absorption of light by the material. The RI of the liquids used to disperse the solids is well known. For isopropyl alcohol, the RI is 1.378 at 20°C. On the other hand, for multi-phasic materials such as binders, RI differs. For example, cement consists of four major components: tri-calcium silicate (RI = 1.72), di-calcium silicate (RI = 1.73), tri-calcium aluminate (RI = 1.71) and tetra-calcium alumino-ferrite (RI = 2.03). Subsequently, choosing an appropriate RI value is difficult. For materials such as binders, the RI values are complex numbers that have both real and imaginary values. For cement, the real RI ranges from 1.5 to 1.8, and the imaginary value is 0.1. Mie theory basically assumes the particles are spherical in nature. The scattered light from spheres of different arbitrary sizes is taken as input along with the RI to determine PSD. In general, particles with sizes ranging from 100 nm to several 100 mm can be analysed by the Mie theory.

On the other hand, for larger particles, the Fraunhofer theory can be used. Moreover, this theory considers the particles as flat and opaque discs. This theory approximates that the size of the particles is larger than the wavelength of the light. Since the diffracted waves alone are considered for the determination of PSD by the Fraunhofer approximation, RI is not required. For materials that are opaque and have high contrast with respect to the liquid medium, the Fraunhofer approximation is limited to a few times greater than the wavelength of the light in order to detect fine-sized particles. Similarly, for transparent particles, the lower end of particle size that LDS can measure using this theory is limited to 25 µm. Usage of this theory requires caution in the case of fine particles. In general, the following guidelines need to be followed in order to choose an appropriate theory for the determination of PSD.

For particles larger than 50 µm in size, this approximation can be used without the need for any optical constants. In the same way, for opaque particles with real RI greater than 1.1, imaginary RI greater than 0.05 and the range of particle size between 1 and 50 µm, Fraunhofer approximation can be used. However, the Mie theory is the best choice if refractive properties are available. If Fraunhofer theory is used for fine particles with real RI less than 1.1 and imaginary RI less than 0.05, then the fine fraction is overestimated during PSD estimation. The refractive properties of the supplementary cementitious materials, such as fly ash, slag and micro silica, are also available in the literature. In general, a real RI value of 1.5 to 1.8 is mentioned for all supplementary cementitious materials. An imaginary RI value of 0.1 is used for fly ash and metakaolin. In the case of microsilica, an imaginary RI value of 0.001 is used.

5.5.5 Errors Affecting Particle Size Measurement

Particle size measurement in LDS is affected by the following variables. They are the solubility of the material, the shape of the particle, the refractive properties of the particle, material heterogeneity, degree of agglomeration, sampling and material concentration during measurement, as well as the mathematical model used in order to analyse the diffracted data.

The mathematical model chosen to process the data plays a vital role in determining the precise PSD. During a PSD analysis, diffracted light is used for the measurement. If the Fraunhofer approximation is being used as a mathematical model, only diffracted light is considered in order to arrive at the diffraction pattern. Subsequently, the reflected and refracted lights are not considered. The angle at which the light reflects and refracts is much larger than the diffraction angle. Since the particle size is inversely proportional to the angle of the beam being received by the detector, accommodating the angle of reflection and refraction results in the detection of an increase in the volume of smaller-sized particles during the analysis. Considerable variations have been observed in PSD analysis due to the decrease in particle size and the ratio of the RI of the sample to that of the medium. Similar to the application of a specific mathematical model, the use of vague refractive properties of the material in the Mie model results in significant errors in the estimation of PSD. Usage of wrong or imprecise RI results in an erroneous assessment of refracted or reflected light waves. Assessment of erroneous waves results in the calculation of different diffraction patterns than required. Consequently, there is an error in the measurement of PSD. An increase in the real part of the RI (opaque) results in the detection of reduced diffracted waves. Reduced diffraction results in an underestimation of the PSD in the finer region. Similar to the real part, an increase in the imaginary RI value (greater than 0.01) results in an increased opacity or translucent nature of the material. Eventually, a steep decrease in the PSD was estimated in the finer regions.

5.5.6 Major Limitations

1. One of the major limitations of using LDS for PSD is the requirement of the refractive properties of the material for the precise determination of the fine fraction of PSD.
2. For particles of size less than 2 μm, the use of the dry technique for the determination of PSD is not recommended as it underestimates the fine fraction.
3. Sampling during the wet technique is a tedious process.

5.6 X-RAY COMPUTED MICRO TOMOGRAPHY

5.6.1 Introduction

X-ray computed tomography (CT) is a technique developed to see inside an object non-destructively and was especially developed to scan human anatomy for medical applications. This technique can provide internal information about the material on different length scales, from metres to nanometres. Currently, this technique is being used in materials and other processing-related industries for understanding and relating micro-structure with the properties of the material. Similar to CT, X-ray micro-computed tomography (μCT) uses X-rays to take projections of an object from different directions. This produces an array of two-dimensional images of the object, also called radiographs. Conventionally, the radiographs are captured on a photographic plate. However, the images can also be acquired digitally. The process of collecting images by X-ray penetration from different directions is called a CT scan. A reconstruction algorithm is used to generate the object from the available array of 2D radiographs. The process of reconstruction generates a 3D object, in greyscale that shows the interior structure of the object. The reconstructed 3D object is referred as '*tomograph*'. The obtained tomograph is processed digitally for further analysis. CT being a non-destructive imaging technique, its application is widely spread. Other than its predominant usage in medicine, CT also has widespread use in forensic analysis, the structural integrity of components that should not be damaged, the examination of delicate samples, etc.

5.6.2 Principle

The contrast of the digital image obtained by the projection depends on the interaction between the X-ray and the object. Based on the RI of the object, intensity and phase changes. Equation 5.32 shows the real and imaginary parts of the RI (n):

$$n = \alpha + i\beta \tag{5.32}$$

where α is the real part that controls the phase shift, imaginary part β controls the absorption (attenuation). Based on the phase shift and attenuation, two components can be obtained. They are phase contrast and attenuation contrast. The attenuation coefficient (μ)is given as shown in equation 5.33:

$$\mu = \frac{4\pi\beta}{\lambda} \tag{5.33}$$

where λ is the wavelength of the X-ray. The attenuation coefficient is a measure of the ease with which the X-ray can penetrate a material.

Several important factors need to be kept in mind while doing a CT scan. A few of them include the size of the object, required temporal resolution, composition, features of interest, spatial resolution, etc. Voxels, or volume elements, are used to represent a 3D image. Choosing an appropriate size for the voxel plays a significant role in characterising features. In order to characterise a feature in μm size, the size of the voxel chosen should be less than the feature size to be characterised; in this case, the voxel should be in nanometres (nm) for proper characterisation. Apparently, the smaller the voxel size, the smaller the size of the object that can be investigated. This is due to the restriction imposed by the detector, followed by the increase in time required for image acquisition, storage and computational reconstruction of acquired projections. In general, increase in the object-to-voxel size ratio increases the parameters related to acquisition and processing.

Contrast between the features of the object is important when choosing an appropriate scanner. Differences in the atomic number of the elements in the object are exploited to get an attenuation contrast in a tomogram. The attenuation coefficient increases with an increase in the atomic number of the elements due to the increased scattering caused by the higher number of electrons in elements with a larger atomic number. For example, a well-known medical application is identifying bone fractures. Apparently, objects that are made up of materials with large differences in their atomic density can be well differentiated by utilising attenuation contrast. On the contrary, phase contrast can be used to distinguish objects made up of materials with low atomic numbers. Energy levels of monochromatic and polychromatic beams can be tailored to get different attenuation contrasts. Optimal beam energy is required to get an ideal attenuation contrast. The lower the beam energy, the lower the signal captured from the object due to the lack of appropriate penetration. Meanwhile, higher beam energy results in poor contrast due to low attenuation. Eventually, the larger the sample size, the higher the optimal X-ray energy required for suitable penetration and contrast. Therefore, the range of energy levels differentiates CT systems, as shown in Table 5.2.

5.6.3 CT Scanner: Configurations and Major Parts

A CT scanner consists of three major components: the X-ray source, the X-ray detector and the stage for placing the sample, as shown in Figure 5.9. The architectural

TABLE 5.2
Energy Range of Different CT Systems

CT System	Energy Range
Nano CT	<1 keV; 5–30 keV
Micro-CT	30–300 keV
Clinical CT	80–140 keV
Industrial CT	>400 keV

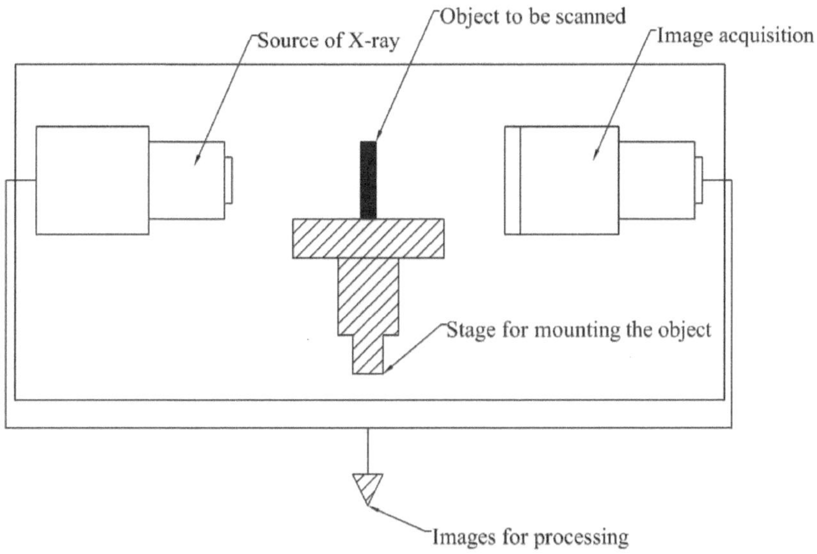

FIGURE 5.9 Schematics of a micro-computed tomography scanner.

design of a CT scanner depends upon the object being imaged, the source and the nature of the X-ray. Based on that, either the object is in a stable position, making it motionless for projection, or the object is moved and the X-ray source is stable.

Synchrotron or X-ray tube acts as a source for X-rays; the X-ray source can be polychromatic or monochromatic based on the type of X-ray source used. Major differences between the X-rays from synchrotron and tube sources are listed in Table 5.3.

Other sub-components in the CT scanner include the condenser and objective lenses, the scintillator, the metal-oxide semi-conductor array, etc. Most of the detectors convert X-rays into visible light with the help of a scintillator. The visible light is converted back into electrons using metal-oxide semi-conductors for digital processing. In order to correct for the variations during the reconstruction of the X-ray projection, an image is also acquired without any sample. This process is called flat field correction. Stability of the instrument and sample is essential for obtaining features of the object without blurring.

TABLE 5.3
Major Differences between a X-ray Tube and Synchrotron

Description	X-ray Tube	Synchrotron
X-ray spectrum	Poly chromatic/mono chromatic	Poly chromatic/mono chromatic
Type of beam system	Cone beam	Parallel beam
Solid angle	Covers significant sample volume	Volume of coverage depends upon the area of the X-ray source
Magnification	Increased or decreased	No changes
Scan time	Minutes to hours	Sub-seconds to minutes

5.6.4 Imaging: Types of Contrast

CT works based on the contrast imaging obtained by the interaction between the X-ray source and the object under imaging. A brief discussion about the types of contrast and the science behind them is discussed in this section.

5.6.4.1 Attenuation or Absorption Contrast

Based on the properties of the materials, encountered during the projection of an object, the X-ray attenuation varies. The attenuation contrast of materials with similar atomic numbers is lower. The variation in the attenuation can be given as shown in equation 5.34, according to Beer–Lambert's law:

$$I = I_0 e^{-\sum_{i=1}^{n} \mu_i x_i} \quad (5.34)$$

where I is the intensity of the X-ray after passing through a material, I_0 is the initial intensity of the X-ray, and μ_i and x_i corresponds to the attenuation coefficient and the path length of the material under consideration. Based on the different parameters, such as the CT scanner, sample and data acquisition, the contrast sensitivity and resolving power of the system are defined. Contrast sensitivity is defined as the level to which minor changes can be detected, and resolving power is defined as the degree to which the notable small changes in close features can be differentiated.

5.6.4.2 Phase Contrast

During a CT scan of an object with different materials, due to the difference in the attenuation coefficient of the materials, attenuation contrast can be used to project and reconstruct the minute differences in the object. However, in some cases, such as soft tissues, the change in the attenuation contrast will be minimal. In those cases, the phase difference in the X-ray can be used to obtain a phase contrast of the object for projection and reconstruction. Phase shift results due to the difference in the real part of the RI (α) as shown in equation 5.32. Deduction of phase shift is done by indirect methods through the measurement of modulated intensity patterns. The simplest measurement technique is propagation-based imaging. In this technique, the X-ray intensity is recorded after it passes through the specimen and space. Waves with different phases interfere, producing crests and troughs at boundaries of the materials. The difference in the boundary pattern is incorporated explicitly during the reconstruction of the projection. Other than propagation-based imaging, interferometric and non-interferometric imaging techniques can also be used. The phase shift cannot be recorded by the detectors; based on the recorded intensity patterns, the phase shift is decoded.

5.6.5 Analysis of Results in the CT Scanner

Collected 2D projections are stacked together and stitched together to obtain a 3D reconstructed image. Analysis is carried out in this image to visualise the specific regions in the volume. This is carried out by assigning a label to every voxel in the volume; the voxels with the same label have the same characteristics (segmentation). Segmentation is followed by the reconstruction of 3D data (volume rendering). The process of capturing and reconstructing the image involves the following steps:

- Slicing of the object
- Acquiring projections at different angles
- Construction of sinograms from the acquired projections
- Back projection of the sinograms to obtain the slices
- Obtaining a filtered back projection slice

The spatial distribution of the attenuation and phase contrast is recovered and reconstructed to form a greyscale image of the object. Radon transform, an integral transformation approach that is used to convert the 2D projections into a 1D profile, is used for recovering and reconstructing the image. The plot of the Radon transform is represented in terms of a sinogram. Based on the beam geometry, each slice is either dependent on or independent of the other slices. In the parallel beam geometry, each slice is independent of the other slices; eventually, the reconstruction of each slice is independent of the other slices. On the other hand, in the case of the cone beam geometry, based on the angle of cone beam projection, each voxel is connected to the other voxel; therefore, the absorption paths are considered 3D. Reconstruction of the 3D absorption image by cone beam geometry can be carried out by back projection algorithms such as Feldkamp, Davis and Kris filtered back projection algorithms. The angle of projection carried out should be less than the minimum voxel size that should be captured. However, considerably fewer projections can also be sufficient for high-quality reconstruction. This can be carried out by iterative reconstruction methods that incorporate prior information about the object being captured; nevertheless, this comes at a higher computational cost.

Matching the precise attenuation coefficient (μ) value is used to match similar material in a tomogram. However, different materials can have the same μ values, therefore, for high accuracy, standard samples should be scanned at similar conditions as the unknown samples to determine the densities with high accuracy by CT. Segmentation and matching of similar constituents in a tomogram is important for quantification of morphological characteristics such as phase fraction, particle size and shape, pore networks, etc. After identifying different voxel values, regions with similar voxel values can be grouped together based on the greyscale threshold, often referred to as the greyscale histogram. A greyscale histogram can be used to differentiate one feature from another in an object, although the simple separation of different features by a greyscale histogram may not be sufficient. In those cases, other methods based on boundary or region-growing approaches are most prominently used.

For any application, the resolution at which the scan needs to be carried out should be fixed based on the smallest details to be observed in the sample. Usually, the resolution is fixed at ten times smaller than the smallest details that can be observed in the sample. The required resolution determines the appropriate CT; either a nano CT or a μCT is used based on the magnification requirement. Based on the path length, the X-ray energy of the beam is determined. For a given sample, the lower the X-ray energy, the higher the absorption contrast between the different materials inside the object; however, the transmission will be lower. The lowest X-ray energy is chosen in such a way that it is sufficient to achieve complete transmission. The exposure time and number of projections are selected appropriately to obtain a sufficient quality of projection. The exposure time and the number of projections are determined based on the scan time and exposure dosage. A synchrotron X-ray source produces a higher X-ray flux; therefore, high temporal resolution can be obtained. On the other hand, X-ray sources from X-ray tubes are better suited for longer-time studies.

5.6.6 Limitations

Every characterisation technique has its own disadvantages. Similarly, X-ray CT also suffers from the following limitations. Based on the different parameters under consideration, following are some of the important parameters that are required to obtain a satisfactory tomogram.

5.6.6.1 Effect of X-ray Dosage

The size of the voxel that needs to be scanned has a significant influence on the dose of X-ray. This generally varies according to the relationship shown in equation 5.35:

$$D \propto \frac{1}{V^4} \tag{5.35}$$

where D is the dosage of X-ray and V is the size of the voxel. Similarly, for high dosage and high flux sources, artefacts and damages are caused during the process of scanning. On the other hand, for X-rays with low energy (less than 30 keV), beam damage and sample heating take place. Sample heating can be significantly reduced by collating the samples using various techniques. However, for applications related to medicine, the effective dose is limited to 200 millisieverts. Increases in X-ray dosage do not have a significant influence on the phase contrast; it decreases less rapidly when compared to attenuation contrast.

5.6.6.2 Artefacts During Imaging

Following are a few of the artefacts that arise during X-ray CT. They are concentric ring artefacts, beam hardening artefacts, limited projection artefacts, and windmill-shaped artefacts that arise when too few projections are used for the reconstruction. Light and dark fringes that are seen at phase boundaries due to the coherence of the beam are called phase contrast artefacts. Other types of artefacts include streaking artefacts and motion artefacts. Artefacts can also arise due to off-centre positioning of the samples during scanning, and cone beam artefacts arise due to the geometrical distortions in cone beam set-ups.

5.6.7 Major Intricacies During the Projection of Construction Materials

Numerous studies are available that use μCT as a technique to categorise concrete and asphalt. Major parameters used for the characterisation are porosity, phase identification, density, damage, failure analysis, in-situ studies, cracks, etc. One of the major differences when projecting a concrete sample when compared to the biological samples is that the atomic weight of the components in the concrete is higher, which leads to the use of X-rays with higher energy. This results in an important artefact related to the projection of concrete samples called the beam hardening effect. The beam hardening effect arises due to the uneven attenuation contrast (which occurs in an object with a material that has a high atomic number) of the material in the object due to polychromatic X-ray radiation. This leads to complete absorption of low-energy X-ray radiation at the middle of the materials with high atomic numbers compared to their periphery. Artefacts due to beam hardening can be reduced by beam filtering. Especially low-energy polychromatic beams can be filtered by absorptive materials such as copper to harden the beam. Other than the beam hardening effect, a type of artefact due to streaking can also take place predominantly. Streaking results in complete absorption of the X-rays by high-density materials in the concrete or by the limitations posed by the detectors.

Other than the artefacts produced during the projection, one of the major limitations of the use of CT for construction material characterisation is the lack of standardised methods. Especially based on the required results, the following information, such as sample type and scan method, should be standardised.

5.7 DYNAMIC VAPOUR SORPTION

5.7.1 Introduction

Uptake of water vapour (moisture) is an inherent property of any material that is also subjected to changes by the prevailing environmental conditions. Understanding the hygroscopic nature of a material is significant in large-scale industrial applications such as food, pharmaceuticals, polymers, the construction industry, etc. In the case of construction, understanding the hygroscopic nature of the material used in the load-bearing and the non-load-bearing structures is substantial for limiting moisture damage. In specific, understanding the water sorption properties of construction materials such as binders, wood, insulation materials, fibres, etc. will help reduce the significant impact of the external environment on limiting the deterioration process. Infusion of moisture into building elements is one of the primary reasons for the onset of various deterioration processes in concrete structures. Variability in the environmental conditions affects the water uptake capacity of building elements such as facades, plasters, etc., which significantly affects the air conditioning loads and quality of the indoor air. Therefore, understanding the hygroscopic nature of solid materials is important in many industries. Dynamic vapour sorption (DVS) is used as a primary technique for measuring water or any solvent uptake by solid materials. Automated gravimetric vapour sorption instruments can be used for measuring the DVS of solid materials. DVS is primarily used to measure water sorption isotherms. Sorption is the process by which a substance gets attached to another substance by physical or chemical means. Uptake of a substance of one state into another substance of a different state by a physical or chemical phenomenon is called absorption. Adhesion of a substance to the surface of a gas, liquid or solid surface is called adsorption. Absorption and adsorption are distinguished by the interaction between the absorbate (adsorbate) and the absorbent (adsorbent). In the case of liquids and gases, the process involves formation of a film of solute on the surface of the solvent. Absorption involves interaction between the substances at a volumetric level. Adsorption is a process in which substances interact at the surface.

5.7.2 Principle

The DVS technique measures changes in mass by varying the vapour surrounding the sample. The vapour is regulated by the controlled mixing of saturated and dry gas streams using controllers. DVS consists of a microbalance with a sensitivity of 0.1 μg, allowing the instrument to detect and measure smaller uptake values. A wide range of organic solvents and water concentrations can be used based on the applications.

Measurement of vapour sorption involves exposing the porous medium (the material to be tested) at a specific vapour pressure. The vapour starts to adsorb/absorb on the surface of the porous media, which leads to a decrease in the vapour pressure of the system. This process goes on until equilibrium is reached between the mass of the adsorbent and the vapour pressure of the system. By repeating the adsorption measurement at different vapour pressures for a porous medium and a vapour (water or an organic solvent) at a constant temperature, the adsorption isotherm of the porous medium can be established.

5.7.3 DVS: Configurations and Major Parts

A DVS consists of two different modules connected to a sample and a reference chamber. The first module controls the microbalance; the gain or loss in weight of the sample with respect to the reference is measured by this module. The second module consists of a reservoir for the control of vapour pressure inside the sample and reference chambers. This particular chamber is used to control the flow of dry and wet gas mixtures that need to be supplied inside the chamber to sustain a specific vapour pressure inside the chamber. The second module is connected to a mass flow controller that controls the flow of dry and moist gas in the area surrounding the sample. The sample and the reference chamber also consist of a real-time humidity vapour

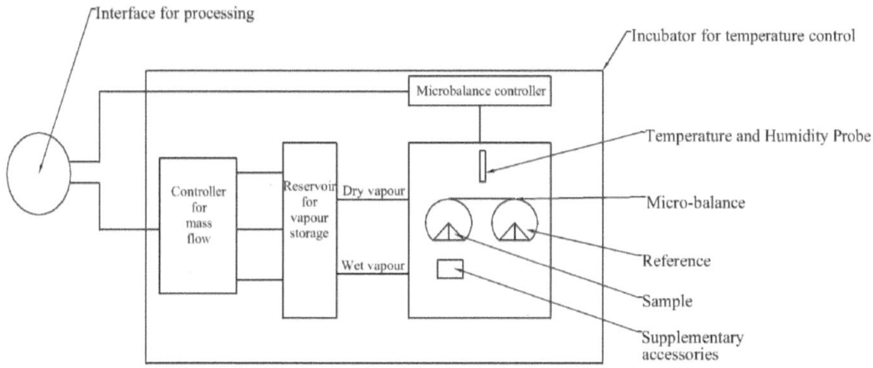

FIGURE 5.10 Schematics of a dynamic vapor sorption instrument.

sensor for recording the humidity inside the chamber; supplementary devices such as a camera, an IR or Raman probe, or both can be fitted to understand the equilibrium process inside the chamber. Modules 1 and 2, along with the reference and sample chambers, are within the temperature-controlled enclosure that can purge automatically to control the temperature inside the chamber. Figure 5.10 shows the schematics of the DVS instrument.

5.7.4 Isotherms

The onset of the adsorption process is induced by different forces between the adsorbent and the adsorbed medium. Different types of isotherms are available in the literature to classify the adsorption process between an adsorbent and an adsorbate. According to Brunauer et al., there are over six types of isotherms to represent the adsorption of adsorbent on adsorbate. For an adsorbent of microporous nature, Type 1 isotherm is widely used. This type of isotherm has a very steep curve with respect to the applied vapour pressure and the volume adsorbed; after the steep increase in the volume adsorbed with respect to the applied vapour pressure, the curve flattens out. Type 2 isotherms are used to represent macro-porous materials (non-porous) when the interaction is between the non-porous adsorbent and an adsorbate. The curve increases very steeply in the initial phase, followed by a gradual increase at the intermediate vapour pressure applied, and finally the volume adsorbed will become asymptotically parallel to the ordinate. Type 3 curves have gradual growth at initial pressure, increases exponentially and become an asymptote with respect to the ordinate. Type 3 curves are typically used for non-porous adsorbents with negligible interactions. For mesoporous adsorbents, type 4 curves are used. This type of curve has a single-step adsorption process. Type 5 curves are used to represent a relatively rare kind of mesoporous adsorbent. Type 6 curves are used to represent adsorbent with multiple steps.

For building materials, the adsorption isotherms are inverted S-type, more similar to the isotherms of type 2. The initial region of the curve describes a relatively fast increase in the moisture adsorbed by the materials; this happens at low relative humidity (RH). The next region in the isotherm shows a gradual, linear increase in the adsorption of the adsorbent on the adsorbate. The final region shows capillary condensation that can be expressed by the Kelvin equation. Similar to adsorption isotherms, an isotherm for desorption can also be obtained by varying the RH. This could indicate an interdependence between the RH and the amount of water desorbed by a material. In general, a hysteresis loop can be identified between a desorption and adsorption isotherm, and the desorption isotherm always lies above the adsorption isotherm. Hysteresis in the isotherm curves is observed due to one of the following reasons. Significant difference in the contact angle before and after desorption due to the presence of the solvent on the surface of the solid, roughness, etc. The second mechanism that is responsible for the hysteresis is due to the alternative wide and narrow parts in the path of sorption and desorption, also called capillary hysteresis. In the narrow parts, the evapouration is significantly slower compared to the wide paths because the partial pressure of the saturated vapour is significantly lower. The above phenomenon increases the capillary condensation process. This leads to slower drying during the desorption process. The shape and position of the adsorption–desorption isotherms are strongly dependent on the temperature. At higher temperatures, the transport of the water molecules is quite easier, therefore, the isotherms that correspond to the higher temperature are lower compared to the isotherms that correspond to the lower temperature. This results in capillary condensation that takes place at a relatively higher RH.

5.7.5 Adsorption Isotherms: Theoretical Approaches

Different theoretical approaches were proposed to describe the experimental data obtained for the porous materials. The Brunauer–Emmett–Teller (BET) isotherm, expressed as shown in equation 5.36, was used to express the surface concentration (C_s).

$$C_s = \frac{C\theta}{(1-\theta)\left[1+(C-1)\theta\right]} C_s^{\text{sat}} \tag{5.36}$$

where, C_s^{sat} is the saturation surface concentration, θ is the coverage, c is the BET constant, which is a function of molar heat of adsorption (H_a) and molar heat of vaporisation (H_c). $c = \exp^{\left[-(H_a+H_c)/RT\right]}$. The applicability of the BET isotherm is limited based on the coverage, which usually $\theta \approx 0.1$. C_s approaches infinity as the coverage reaches 1. Certain modified equations were proposed by Brunauer–Skalny–Bodor (BSB) and Brunauer–Deming–Deming–Teller (BDDT) isotherms, as shown in equations 5.37 and 5.38, with three parameters rather than two. The BSB isotherm is given according to equation 5.37:

$$C_s = \frac{Ck\theta}{(1-k\theta)\left[1+(C-1)k\theta\right]} C_s^{\text{sat}} \tag{5.37}$$

The third fitting parameter k should be less than 1.0 and the adsorbed layers are considered to be identical other than the fact that the state of the adsorbed layers is not in a pure liquid form. In the case of the BDDT isotherm, an additional parameter n is included that represents the number of adsorbed layers. The BDDT isotherm is shown in equation 5.38:

$$C_s = \frac{c\theta}{(1-\theta)} \frac{1-(n+1)\theta^n + n\theta^{n+1}}{1+(c-1)\theta - c\theta^{n+1}} C_s^{\text{sat}} \tag{5.38}$$

Another model was proposed after the Frenkel–Halsey–Hill (FHH) isotherm, as shown in equation 5.39. This model was proposed to accommodate high θ values.

$$C_s = \left(-\frac{K}{ln\theta}\right)^{1/\sigma} C_s^{\text{sat}} \tag{5.39}$$

5.7.6 Measurement and Details

After drying the samples, they were placed in the sample chamber of the DVS instrument. Based on the requirements, the temperature of the chamber can be set from 5°C to 60°C. Once the temperature was set, the sample in the chamber was exposed to different partial pressures. The RH is varied according to the following percentages: 0%, 10%, 20%, 30%, 40%, 50%, 60%, 70%, 80%, 90% and 98%. The DVS instrument is run in mass variation over time variation mode (dm/dt) until equilibrium is reached at each RH. The values of dm/dt are fixed based on the requirement. Fixing a specific value is required for the DVS instrument to determine the equilibrium position reached by the sample at each RH. The criteria fixed are used to complete a specific RH step and advance to the subsequent steps for the measurement. When the rate of change of mass goes below the fixed criteria during the fixed time period, the measurement of mass change will proceed to the next set level. A typical measurement curve is shown in Figure 5.11. These curves are used to identify

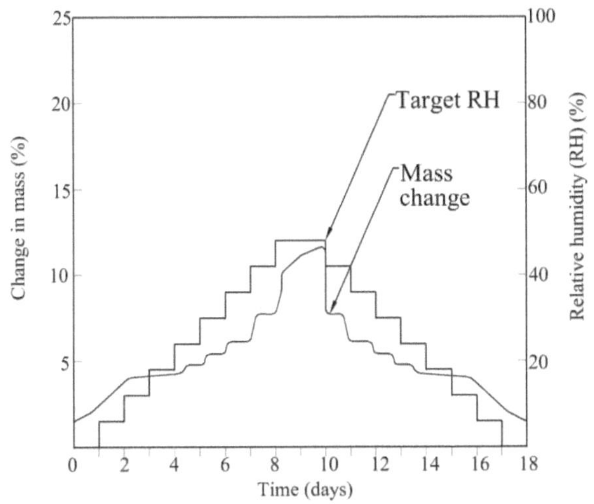

FIGURE 5.11 A typical measurement curve showing the variation in the mass with respect to time for various target relative humidity at a constant temperature.

the water uptake rate and the coefficient of water diffusion. Sorption and desorption isotherms are calculated from the measured equilibrium mass values at the humidity step. The hysteresis loop between the sorption and desorption isotherms and the shape of the isotherms are used to identify the sorption mechanism and the porosity of the sample.

MULTIPLE CHOICE QUESTIONS

1. Based on which of the following molecular process, IR spectroscopy is designed __________
   a. Rearrangement of nuclear particles
   b. Molecular rotation
   c. Molecular vibration
   d. All the above
2. The rearrangement of nuclear particles happens in one of the following region__________
   a. X-ray
   b. Gamma ray
   c. Microwave
   d. Infrared
3. A band of wavelength produced by the refraction of white light to different degrees is called as __________
   a. Emission
   b. Absorption
   c. Spectrum
   d. All the above
4. The occurrence of spectral lines during the emission or the absorption process are __________ for liquid and gas.
   a. Very sharp spectral lines
   b. Less sharp spectral lines
   c. Both of the above
   d. None of the above
5. Doppler broadening effect is the most predominant in __________
   a. Liquid, gas and solid
   b. Liquid and solid
   c. Liquid and gas
   d. None of the above
6. Effect of the uncertainty principle is not negligible in case of which of the following molecular process __________
   a. Excited electron in spin state
   b. Nuclear particle rearrangement
   c. Rearrangement of inner electrons
   d. Vibration and rotation of a molecule
7. In IR spectroscopy, C–S–H gel can be identified at __________
   a. 938–883 cm^{-1}
   b. 995–900 cm^{-1}

   c. 1,000–950 cm^{-1}
   d. None of the above
8. The presence of portlandite can be identified by a single peak at __________
   a. 3,640 cm^{-1}
   b. 1,115 cm^{-1}
   c. 4,250 cm^{-1}
   d. 915 cm^{-1}
9. Energy and size of the orbital are governed by __________
   a. Azimuthal quantum number
   b. Spin quantum number
   c. Magnetic quantum number
   d. Principal quantum number
10. The principal quantum number takes any integral values from __________
    a. Zero to infinity
    b. One to infinity
    c. ± azimuthal quantum number (l), ±(l–1), __________0
    d. –infinity to +infinity
11. The azimuthal or the orbital quantum number takes integral values from __________
    a. Zero to infinity
    b. One to infinity
    c. ± principal quantum number (n), ± (n–1), __________0
    d. –infinity to +infinity
12. Spin angular momentum is due to __________
    a. Rotation of sub-atomic particles about its own axis
    b. Revolution of sub-atomic particles
    c. Revolution of the electron around the nucleus
    d. All the above
13. The spin angular momentum is a vector quantity that takes a spin value of __________
    a. ±½
    b. $\frac{\sqrt{3}}{2}\frac{h}{2\pi}$
    c. $s = \sqrt{s(1+s)}\frac{h}{2\pi}$
    d. All the above
14. If the deuterium has spin in the opposite direction, then the total spin value of deuterium is __________
    a. 1
    b. 0
    c. ½
    d. –1

15. If a nucleus has an odd number of protons and even a number of neutrons, then the spin value of the nucleus will be __________
    a. Zero
    b. Integral value
    c. Half-integral value
    d. All the above
16. All the components in elements are in unexcited state then the energy state is called as __________
    a. Degenerate
    b. Unlike state
    c. Dis-similar state
    d. All the above
17. Energy is emitted or absorbed in the form of radiation at a specific frequency, if a nuclear spin transfers from one to another orientation. Which of the following is correct?
    a. The magnetic field applied is directly proportional to the radiation frequency
    b. The magnetic field applied is inversely proportional to the radiation frequency
    c. Both a and b
    d. None of the above
18. The sharing of excess energy with the adjacent nuclei or the surrounding is called as __________
    a. Relaxation time
    b. Relaxation process
    c. Resonance
    d. All the above
19. Consider an example where a hydrogen is paired with an oxygen and a hydrogen is paired with a carbon. In which case the shielding effect is higher for hydrogen atoms?
    a. Hydrogen paired with oxygen
    b. Hydrogen paired with the carbon
    c. In both cases, it remains the same
    d. None of the above
20. What happens to the resonance frequency of a hydrogen paired with a highly electro-negative system?
    a. Resonance takes place at a higher value
    b. Resonance takes place at a lower value
    c. No change in the resonance frequency
    d. None of the above
21. __________ is based on the influence of the magnetic effect of one nucleus on the other.
    a. Coupling constant
    b. Chemical shift
    c. Larmor frequency
    d. Relaxation time

22. In UV–VIS, the light absorbed by the sample is given by __________
    a. $\log_{10}\frac{I_0}{I}$
    b. $\log_{10}\frac{I}{I_0}$
    c. $\log\frac{T}{T(1-\alpha)+\alpha}$
    d. None of the above
23. What makes mercury to be used as an intruding agent in MIP?
    a. High surface tension and low contact angle
    b. Low surface tension and high contact angle
    c. High surface tension and high contact angle
    d. Low surface tension and low contact angle
24. Higher is the contact angle between mercury and sample under examination __________
    a. Higher is the pressure required for mercury to intrude the surface
    b. No change in the pressure required
    c. Lower is the pressure required for mercury to intrude the surface
    d. None of the above
25. Pressure applied and the radius of the capillary are __________
    a. Directly proportional
    b. Inversely proportional
    c. No direct relationship
    d. None of the above
26. Contact angle used during the examination of a cement-based system __________
    a. 90
    b. 10
    c. 140
    d. 0
27. The step method is considered as the best method in determining the pore size distribution because __________
    a. Delicateness in performing
    b. An equilibrium that can be achieved
    c. The sample will remain intact
    d. None of the above
28. Increasing the input value of contact angle results in __________ pore size.
    a. Increased
    b. Decreased
    c. Same
    d. Increased and decreased

29. In general, bigger sized samples are preferred because __________
    a. To avoid local heterogeneities
    b. To avoid the ink-bottle effect
    c. To achieve equilibrium
    d. None of the above

30. Which of the following method/ methods of cement sample preparation results in larger pore size distribution in MIP?
    a. Solvent exchange method
    b. Oven drying
    c. Freeze drying
    d. Vacuum drying
31. On curing similar cement-based samples underwater and air, what happens to pore size distribution?
    a. Increases in case of underwater curing compared to open cured samples
    b. Decreases in case of underwater curing compared to open cured samples
    c. Both a and b
    d. None of the above
32. MIP measures __________
    a. Pore size distribution
    b. Entry points of the pores
    c. Small pores
    d. Larger pores
33. MIP has one of the following major limitations __________
    a. It can intrude only unconnected pores
    b. It can intrude through all kinds of pores
    c. It can intrude only connected pores
    d. None of the above
34. Measurement through laser diffraction (LD) requires __________
    a. Optical properties of the materials
    b. Physical properties of the materials
    c. Chemical properties of the materials
    d. Surface properties of the materials
35. Refractive index (RI) is not required for __________ scattering model.
    a. Fraunhofer model
    b. Mie model
    c. Both a and b
    d. None of the above
36. Which of the following phenomena are independent of composition and optical properties of the material __________
    a. Reflection
    b. Refraction
    c. Absorption
    d. Diffraction

37. The Refractive index (RI) of the material is divided into real part and imaginary part. The imaginary part of the RI depends on which of the following factors?
    a. Scattering
    b. Reflection
    c. Refraction
    d. Absorption
38. RI of IPA is __________
    a. 1.972; 0.1.
    b. 1.378; 0.0.
    c. 1.378; 0.1.
    d. 1.972; 0.0.
39. In the case of Fraunhofer approximation, only __________ light is considered in order to arrive at diffraction pattern.
    a. Scattered
    b. Reflected
    c. Absorbed
    d. Diffracted
40. The size of the particle is __________ to the angle of the beam.
    a. Directly proportional
    b. Inversely proportional
    c. In no relationship
    d. None of the above

1 c	2 b	3 c	4 b	5 c	6 a	7 c	8 a	9 d	10b
11a	12a	13a	14b	15c	16a	17a	18b	19b	20a
21a	22a	23c	24a	25b	26c	27b	28a	29a	30b
31b	32b	33d	34a	35a	36d	37d	38b	39d	40b

Index